Of
Knowledge
and
Faith

PHILIP L. TAN

Of Knowledge and Faith

Does God Exist?

Copyright © 2020 by Philip Tan

All rights reserved. No part of this publication may be reproduced, stored in a retrieval system, or transmitted, in any form or by any means, except as may be expressly permitted "by the 1976 Copyright Act" or by Philip Tan in writing.

ISBN: 978-1-7351344-0-6 (paperback)
978-1-7351344-1-3 (ebook)

Library of Congress Control Number: 2020914647

Book design by Wordzworth
www.wordzworth.com

Contents

INTRODUCTION		IX
PROLOGUE		XI
CHAPTER 1	THE ASCENT OF MAN	1
CHAPTER 2	THE RELIGIONS OF MAN	9
CHAPTER 3	IN MAN'S IMAGE AND LIKENESS	13
CHAPTER 4	APOTHEOSIS	17
CHAPTER 5	APOTHEOSIS II	23
CHAPTER 6	THE WORSHIP OF MAN	27
CHAPTER 7	THE NATURE OF MAN	31
CHAPTER 8	EUREKA!	37
CHAPTER 9	SAPERE AUDE!	43
CHAPTER 10	PROMETHEUS UNBOUND	49
CHAPTER 11	AUDEAMUS!	55
CHAPTER 12	ON MORALITY	65
CHAPTER 13	CONCLUSION	69
ACKNOWLEDGEMENTS		75
NOTES		77
BIBLIOGRAPHY		85
INDEX		95

List of Tables

Table 1-1	Comparative DNA: Similarity between Select Organisms and Humans	2
Table 2-1	Comparative DNA: Similarity between Neanderthals and Humans	10
Table 2-2	Evolution of Religions with Human Settlements, 35,000 B.C.E. to 2000 C.E.	11
Table 3-1	Egyptian Gods and Goddesses	14
Table 3-2	Greek Gods and Goddesses	16
Table 4-1	Covenants between God and the Jews – Jewish Bible	19
Table 5-1	Covenants between Jesus and Humanity – New Testament	25
Table 6-1	Vital Statistics – the Middle Ages (500 C.E.–1500 C.E.)	30
Table 8-1	Significant Inventions and Scientific Discoveries of the Seventeenth Century	40
Table 9-1	Eighteenth-Century Philosophers and Their Key Ideas	46
Table 10-1	Scientific Discoveries and Historic Inventions from the First Century to the Twentieth Century	54
Table 11-1	Selected Scientific Discoveries and Inventions of the Twentieth Century	61
Table 13-1	Differences between Science and Judeo-Christianity	73

List of Illustrations

Image 1-1	Divergence of the Family Tree between Hominins and Great Apes	3
Image 1-2	Location of Hominin Fossil Deposits	4
Image 1-3	History of Hominins	5
Image 1-4	Comparison between Ape and Hominin Skeletons	6
Image 3-1	Scale model of the West Pediment of the Parthenon in Athens	15
Image 7-1	Renaissance Man: Leonardo da Vinci (1452–1519)	36
Image 11-1	Tree of Life Based on the Sequenced Genome	62
Image 11-2	Graphical Representation of the Milky Way	63

Introduction

*"The key to wisdom is this—constant and frequent questioning ...
for by doubting we are led to question and by
questioning we arrive at the truth."* [1]

— PETER ABELARD —

Does God exist? This is the nagging and underlying question on which this book is predicated. Though theology assumes the obvious, everything in it literally begs the same question. However, a run through history is the most likely source of a convincing, albeit not reassuring, answer based on reliable evidence.

The "God" referred to in this question pertains only to the Judeo-Christian tradition, whose rich heritage and readily available body of historical materials and publications have sustained endless interest in it. It is, however, important to have an understanding of pre-Judeo-Christian religious developments before delving into this rich heritage. Hence, Chapter 1 deals with the origin of man, while Chapter 2 relates to the evolution of religion. Chapter 3 is a religious snapshot of Egyptian and Greek civilizations, which are both well known and seminal in Western thought. Chapters 4 and 5 cover Judeo-Christianity, while Chapters 6–11 trace the journey of Christianity from the second century C.E. to the twentieth century. Chapter 12 takes on morality, while the final chapter presents the conclusions expressed in hypotheses.

This book is not a sweeping account of Western religious history. It is an account given from the perspective of a Western mind-set that seeks to serve as a sound prelude for further investigation. The focus on the essentials, which were synthesized from various sources, will hopefully add to the body of work available to students of history and seekers of the truth.

Prologue

The following is an excerpt from Christopher Hitchens' speech at the "Festival of Dangerous Ideas," held in the Sydney Opera House, Australia, in October 2009:[1]

"There is a difference of opinion about how long our species in general has been on the planet. It's a flash of a second in evolutionary time, which (Richard) Dawkins thinks might be about as long as 250,000 years … make it a quarter of a million. Francis Collins, the man who did the DNA decoding—the Human Genome Project—who is, by the way, another Christian scientist, or rather a scientist who is a Christian, will not make that mistake again. You know what Mark Twain said about the work of Mary Baker Eddy, Christian Science writer, the founder of it, her books—chloroform in print. This is a digression.

Francis Collins says at least 100,000 years. We can show we've been around that long. Quite a long time. I'll take a one hundred; never mind, I don't need a quarter million for my point. Make it 100,000. A hundred thousand years people have been … our species have been around on this speck; born … usually dying actually, a great number of them in childbirth, they don't get beyond being born. For the first eighty or so, ninety or so thousand years, nearly one hundred, not living more than twenty-five or thirty years old at the most and then probably dying of their teeth if they were lucky. Or the other needless mammalian things that show us that we bear the stamp, as (Charles) Darwin put it, of our lowly origin ... the appendix we don't need anymore, innumerable other shortcomings of our design. We're designed to live in the savannah we've escaped from. Terrible disease, suffering, misery, malnutrition. And fear. Where did the earthquakes come from? Why is there an eclipse? What are the shooting stars doing? And awful cults of sacrifice trying to ward off what are in fact natural events, and war, and rape and the kidnap of other peoples and the enslavement of ... and all of this goes on.

Gradually, gradually inching up to the point where you can brew beer (a breakthrough in my view), domesticate animals, separate one kind of corn from another. Millimetric progress, terrible struggle, sacrifice, pain, misery but, above all, fear and ignorance. And you have to believe this if you believe in monotheism: For the first 97–98,000 of this, heaven watches with indifference. Oh, there they go again. That whole civilization just died out. Well, what are you going to do? They're raping each other again. They're poisoning ... They think the other tribe has poisoned the well so they're going to kill all their children. Just watch all that.

Three thousand years ago at the most, it was decided: No, we have to intervene now. You have to believe it. You have to believe it. And the revelation is, must be ... must be personal ... must appear. So, we'll pick the most backward, the most barbaric, the most illiterate, the most superstitious, the most savage people we can find in the most stony area of the world. We won't appear to the Chinese who can already read. We would appear in the Indus Valley where they know a thing or two. And they're already, you know, they're very far advanced. No, we will appear to this brutal, enslaved, hopeless, superstitious crowd and will force them to cut their way through every and all of their neighbors with slaughter, genocide, and racism and settle on the only part of the Middle East where there's no oil. And all subsequent revelations occur in this same district. And without this, we wouldn't know right from wrong."

1
The Ascent of Man

"Man with all his noble qualities, with sympathy which feels for the most debased, with benevolence which extends not only to other men but to the humblest living creature, with his god-like intellect which has penetrated into the movements and constitution of the solar system—with all these exalted powers—Man still bears in his bodily frame the indelible stamp of his lowly origin." [1]

– CHARLES DARWIN –

DNA—LIFE'S COMMON THREAD

Life forms, from the simplest, such as bacteria and fungi, to the most complex, such as plants, animals, and humans, are made up of cells; within those cells lies the chemical stuff known as DNA (deoxyribonucleic acid) that, according to *Encyclopedia Britannica*, "codes genetic information for the transmission of inherited traits and is found in all prokaryotic and eukaryotic cells and in many viruses." It is in essence what makes a rose a rose, an oak an oak, a cat a cat. All life forms have DNA as a common denominator; therefore, all life is related to all other life through this common heritage.

With the apparent discovery of DNA in 1869 and the subsequent identification of its structure in 1953, scientists opened the book of life whereby coded information in the form of genes is literally read, identified, and compared in organisms.[2]

These investigations either preceded or followed the Human Genome Project, which was completed in 2003 and identified and sequenced the 22,300 protein-coding genes in human DNA and their 3.3 billion chemical base pairs.[3] Table 1-1 presents some comparisons between the human genome and that of other organisms.

Table 1-1 Comparative DNA: Similarity between Select Organisms and Humans

	% Similarity to Human DNA
Chimpanzee	98.8%
Bonobo	98.8%
Gorilla	97.7%
Orangutan	96.9%
Old World Monkey	93.0%
Cat	90.0%
Mouse	85.0%
Dog	84.0%
Cow	80.0%
Chicken	65.0%

	% Similarity to Human DNA
Fruit Fly	61.0%
Banana	41.0%

The human species *Homo sapiens* is classified in the family of the great apes, together with the orangutan, gorilla, and chimpanzee.[4] *Homo sapiens* is the closest relative of the chimpanzee and the bonobo, but that is not all. Humans share more of their DNA with mice, at 85%, than with a chicken at 65% (we even share genes with bananas!).

In addition, the similarity of the human genome with that of the chimpanzee suggests that at one point, millions of years ago, the two species had a common ancestor. About seven million years ago, the early hominins diverged from the African great apes. See Image 1-1.

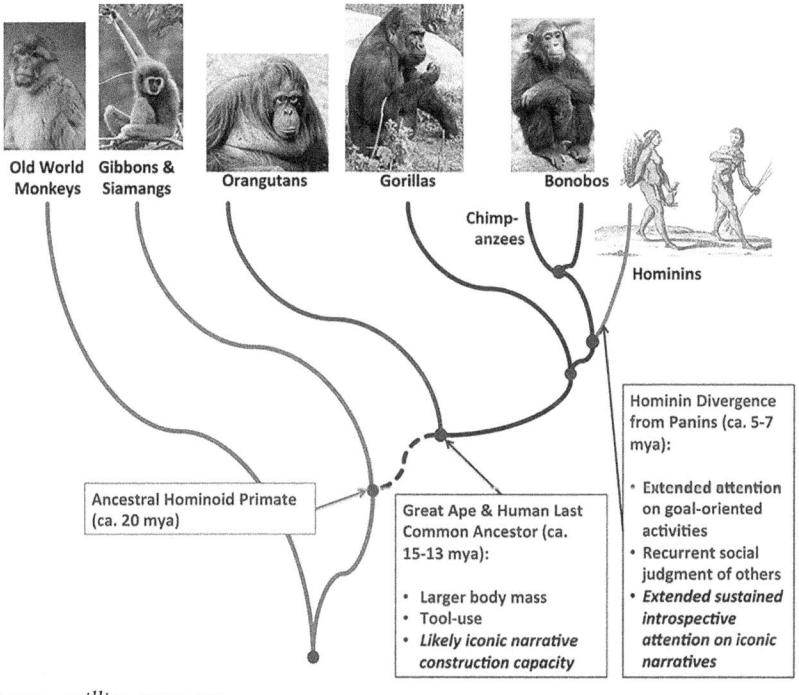

*mya – million years ago

Image 1-1 Divergence of the Family Tree between Hominins and Great Apes[5]

CHARLES DARWIN VINDICATED

When Charles Darwin published *The Descent of Man*, in 1871, he postulated, among other things, that humans were descended from the great apes. He presciently suggested that researchers look for man's progenitors in Africa, where the great apes still abounded. He was soon proven correct—beginning in the late nineteenth century, evidence of fossil records validated his theory (Image 1-2). Moreover, the study of DNA in the twentieth century only cemented Darwin's claim to have established indisputable fact.

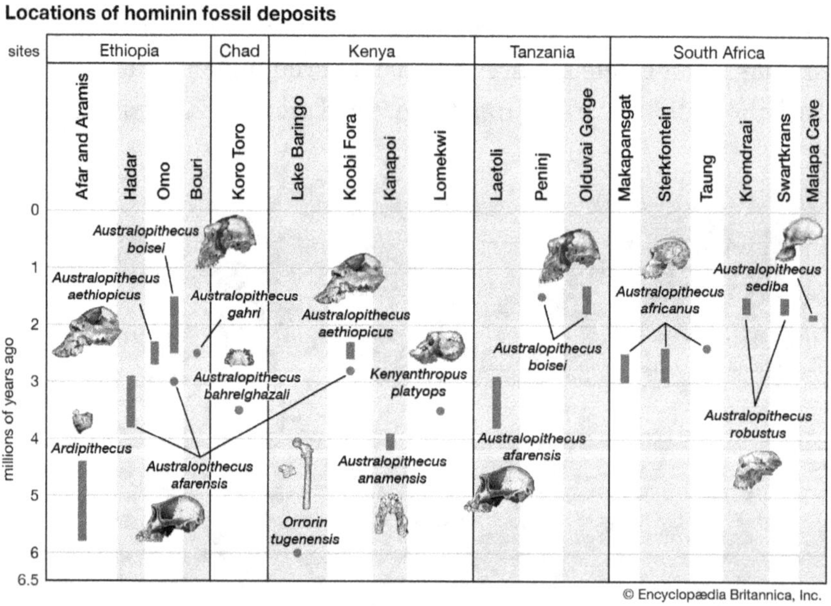

Image 1-2 Location of Hominin Fossil Deposits[6]

APE VERSUS HOMININ

In 1891, on the island of Sumatra in present-day Indonesia, the Dutch anthropologist Eugene Dubois and his team discovered a femur bone and skull cap that were identified as the first human fossil remains known to modern science, believed to belong to a primitive hominin species, later given the name *Homo erectus*. Other paleontologists discovered many more premodern human remains in Africa as well as around the globe.[7] Image 1-3 shows the treasures they discovered.

Hominins are divided into two genera: *Homo* and *Australopithecus*. It is clear that the two genera coexisted in many ways over millions of years until their extinctions. *Homo sapiens* is the only remaining hominin species alive.

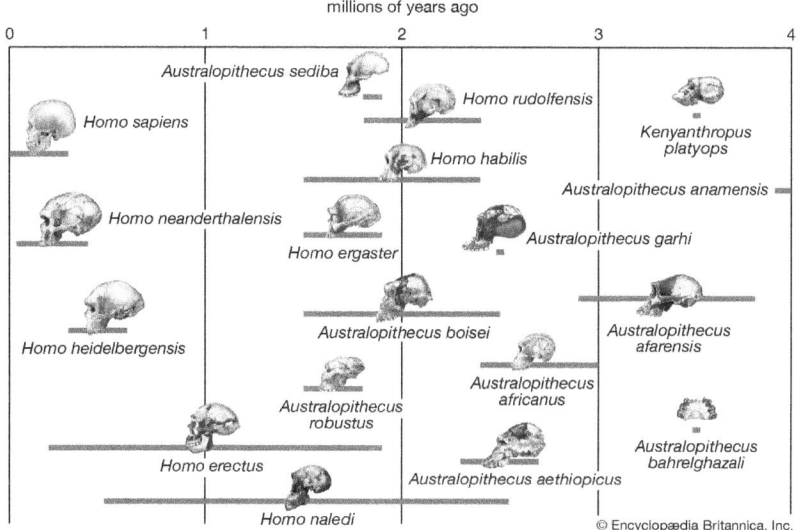

Image 1-3 History of Hominins[8]

How do scientists determine whether a set of fossil remains belong to early hominins or just plain apes? They are classified using a series of criteria, among which are the following:

1 Bipedalism, or the ability to consistently stand upright while walking or running
2 Skeletal differences
3 Smaller canines, thick tooth enamel, fewer molars
4 The use of handheld stone tools

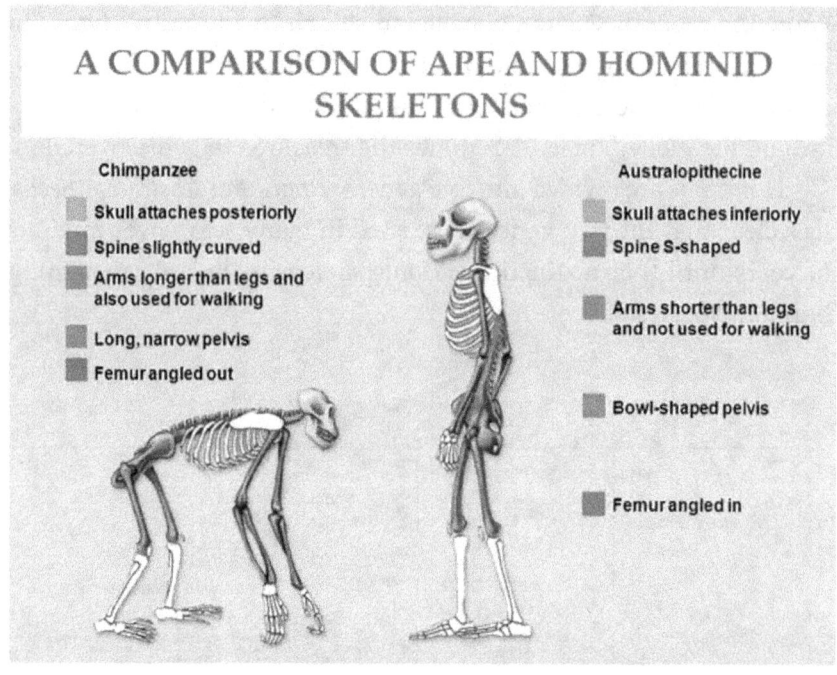

Image 1-4 Comparison between Ape and Hominin Skeletons[9]

Homo sapiens: THE LAST HOMININ STANDING

The fossil record shows that *Homo sapiens* emerged in eastern Africa some 200,000 to 150,000 years ago. *Homo erectus* and the Neanderthals (*Homo neanderthalensis*), who preceded the arrival of *Homo sapiens*, were also his contemporaries.

Until 70,000 years ago, *sapiens* remained largely confined to eastern Africa before it pushed out to the territories occupied by *erectus* (Africa, Europe, and Asia) and *neanderthalensis* (Europe, the Middle East, and Asia). Before long (35,000 years later), *erectus* had disappeared, and *neanderthalensis* followed soon thereafter.

The size of the brain of *sapiens* was likely the key to its domination. As Niles Eldredge wrote, "The high foreheads of our own species, *Homo sapiens*, mark the presence of an expanded mass of the brain's cerebrum: We are literally far more cerebral than our Neanderthal cousins were."[10] Nevertheless, it is not known what that big brain became used for between 70,000 and 35,000 years ago that pushed *sapiens* to venture out of Africa, although it seems likely that during that period, he became able to harness its innate skills—cognitive, linguistic, creative, social, organizational, and cultural—to adapt to, or even shape, the environment.

2

The Religions of Man

"Man appeared and turned the course of this evolution from an indefinite march of physical aggrandizement to a freedom of a more subtle perfection. This has made possible his progress to become unlimited, and has enabled him to realize the boundless in his power." [1]

— RABINDRANATH TAGORE —

THE GREAT LEAP FORWARD

For about 130,000 years, before *Homo sapiens* emerged from Africa for the first time, he and the Neanderthals were little different in their behavior and culture, except for their physical appearance and brain size. Both used the same stone tools, were hunter-gatherers, and employed fire for cooking; both were nomads who probably gathered in bands or tribes.

The preliminary results of sequencing the Neanderthal genome and comparing it to those of modern humans (see Table 2-1), completed in 2010, showed that the two are nearly identical.[2]

Table 2-1 Comparative DNA: Similarity between Neanderthals and Humans

	% Similarity to sapiens DNA
Neanderthal	99.7%

Because *sapiens* had a smaller brain than the Neanderthals (1,400 cc vs. 1,600 cc), it is believed that they may have had no comparative advantage until a mutation occurred 70,000 years ago to bestow on *sapiens* the ability for speech and language.[3] The anatomical additions of a skull base, the hyoid bone that supports the tongue, and the lower larynx enabled *sapiens* to communicate more clearly among themselves,[4] thus enabling closer cooperation and providing an enriched shared culture.

HISTORICAL RELIGIONS

Although speech and language were evolving, *sapiens* was aware that it did not have control over its surroundings, other animals, the weather, or nature in general; instead, humans totally depended upon all of these for their lives. They likely believed that these were all alive as well. Animism refers to the belief that all things, living and nonliving, including nature itself, have spirits.

Mythology is the set of stories passed on from one generation to the next about how the world came to be; how nature influences the life of humans and animals; and the undeniable link between the visible and the invisible worlds, between man and gods, and between natural and supernatural forces. In this understanding, man had the role of propitiating those higher, invisible powers that were thought to control nature and, hence, his own fate in this world and the afterlife, using rites and rituals.

Language enabled ideas and information to be shared and stored in an oral tradition among many, not just within a tribe.

After small-scale agriculture, often called horticulture or subsistence farming, began in earnest 10,000 years ago, *sapiens* began settling down in villages.[5] Once humans found out how to domesticate crops and livestock for food, living in towns began. As people gathered in villages and towns, one of the shared experiences in a community was religious worship. Table 2-2 gives a summary of how religion evolved as humans gathered into ever larger polities.[6]

Table 2-2 Evolution of Religions with Human Settlements, 35,000 B.C.E. to 2000 C.E.

Period - B.C.E.	Settlement	Religion	Evidence
35,000–10,000	Nomadic	Animism	Burial of the dead
			Grave gifts
			Animal sacrifice
			Fetish cults
			Cave carvings
			Cave paintings
			Shamanism
10,000–3,000	Village	Mythology	Temples
	Town		Monuments
			Megalithic grave markers
			Ancestor cults
			Human sacrifice
			Priesthood

Period - B.C.E.	Settlement	Religion	Evidence
3,000–1,500	City	Polytheism	Worship of deities Mummification Festivals
1,500–0 C.E.	City-state	Polytheism	As above
0–1600 C.E.	Kingdom	Monotheism	Cathedrals/churches Sculptures Paintings
1600–2000 C.E.	Nation-state	Monotheism	As above

Using the functions of its well-developed frontal lobe for abstract thinking, problem solving, planning, judgment, emotions, intelligence, and self-consciousness, *sapiens* was on its way to dominating the world around it through sheer manipulation. As Lewis Mumford put it,

> As far as the present record stands, grain cultivation, the plow, the potter's wheel, the sailboat, the draw loom, copper metallurgy, abstract mathematics, exact astronomical observation, the calendar, writing and other modes of intelligible discourse in permanent form, all came into existence at roughly the same time, around 3,000 B.C.E. give or take a few centuries. The most ancient urban remains now known, except Jericho, date from that period. This constituted a singular technological expansion of human power whose only parallel is the change that has taken place in our own time. In both cases men, suddenly exalted, behaved like gods: but with little sense of their latent human limitations and infirmities, or of the neurotic and criminal natures often freely projected upon their deities.[7]

3

In Man's Image and Likeness

"Ethiopians imagine their gods as black and snub-nosed; Thracians blue-eyed and red-haired. But if horses or lions had hands, or could draw and fashion works as men do, horses would draw the gods shaped like horses and lions like lions, making the gods resemble themselves." [1]

— XENOPHANES OF COLOPHON —

"Men create gods after their own image, not only with regard to their form but with regard to their mode of life." [2]

— ARISTOTLE —

ANCIENT EGYPT (3150–332 B.C.E.)

Polytheism is the veneration of many deities. In Ancient Egypt, whose civilization endured for more than 2,800 years, hundreds of deities were worshiped. Egypt was divided into between thirty-five and forty provinces, and each province had its own local gods, as did each major city.[3] This variety reflected man's insatiable needs and wants. However, the precise supernatural powers assigned to given entities were various, with each having an area of specialization.

Religion was at the core of Egyptian civilization. Will Durant wrote that "We find it in every stage and form from totemism to theology; we see its influence in literature, government, art—in everything except morality."[4] Most of the principal gods were depicted as humans with the head of an animal or were given animal characteristics. Their images were hewn in stone or painted on walls. The pharaoh, too, was considered the personification of a god. His divinity was supported by an all-male priesthood that exhibited unquestioning loyalty.[5]

A prominent feature of the religion of Egypt, at least among the rich and powerful, was its preoccupation with funeral preparation (mummification) and attendant offerings for the dead, along with elaborate, detailed paintings of scenes of the afterlife on tomb walls.[6] Table 3-1 gives a list of the principal Egyptian gods.

Table 3-1 Egyptian Gods and Goddesses

Egyptian Gods and Goddesses	
Amun, Ra or Amun-Ra	Sun god, creator god, protector of kings, father of mankind
Anubis	God of the dead
Hathor	Goddess of women, love, and pleasure
Horus	The sky god
Isis	Mother goddess and goddess of harvest/food
Khepra	A form of Ra
Khnum	God of water, creator of the gods and mankind

Egyptian Gods and Goddesses	
Mut	Queen of the gods, divine mother
Osiris	God of vegetation and of the underworld
Ptah	Creator god and god of handicrafts
Seth	God of darkness, chaos, and storms
Thoth	Moon god and god of books, wisdom, and letters

ANCIENT GREECE (776–323 B.C.E.)

The Ancient Greeks had many gods, described in two epics of Homer, the *Iliad* and *Odyssey*, as well as Hesiod's *Theogony*. Beyond the supernatural attributes they possessed, they were all portrayed as human in form and character, mirroring human vices and virtues, behavior, and temperament. The Ancient Greeks appeared to have created their gods in their own image and likeness. Today, we call this tendency anthropomorphism. The Greeks sculpted the images of their gods in marble.

Image 3-1 Scale model of the West Pediment of the Parthenon in Athens. In the center Athena and Poseidon appear, surrounded on both sides by minor deities and other mythological figures.[7]

Greek religious practice included libations, which involved pouring out wine or olive oil as an offering to the gods at shrines and in homes; animal sacrifices given in front of the altar while singing hymns and praying; votive offerings of food, drinks, and precious objects at temples or shrines; and festivals held within a city or regionally, such as the Olympic Games and the other three Panhellenic games. The Ancient Greeks also flocked to the Temple of Apollo at Delphi to seek counsel from a female oracle, and they also took direction from the gods through dreams, seers, drawing lots, or burnt offerings.[8]

Table 3-2 Greek Gods and Goddesses

Greek Gods and Goddesses	
Aphrodite	Goddess of love, beauty, and fertility
Apollo	God of archery, music, medicine and the sun
Ares	God of war
Artemis	Goddess of hunting and childbirth
Athena	Goddess of wisdom and war
Demeter	Goddess of harvests
Dionysius	God of wine, fruit, and crops
Hephaestus	God of fire and craftsmen
Hera	Wife of Zeus, patroness of marriage and women
Hermes	God of merchants and traders and messenger of the gods
Hestia	Goddess of the hearth, home, and family
Poseidon	God of the sea and the cause of earthquakes
Zeus	Father or supreme god and god of the sky and weather

4

Apotheosis

"What are men? Mortal gods.
What are gods? Immortal men." [1]

— HERACLITUS —

"Man in his arrogance thinks himself a great work
worthy of the interposition of a deity." [2]

— CARL SAGAN —

ALEXANDER III

Alexander the Great died in 323 B.C.E. in Babylon, after defeating the Persian Empire and expanding Greek civilization into the Near East and Asia through his immense conquests, accomplished over the span of only a decade. His exploits were so uncanny that Alexander was deified by at least one of his successors. When he was a child, Alexander felt a resonance with his mother's claim to be descended from Achilles and his father's claim to be descended from Heracles.

He visited oracles twice: first at Delphi, before he set out to conquer Asia, and then at Siwa in Libya, a year into his campaign. At Delphi, where the priestess declared to him "You are invincible, my son," Alexander was emboldened. At Siwa, the priest told him that he truly was the son of Zeus and would rule the world. After this, Alexander believed his true father to be Zeus.[3]

ROMAN EMPERORS

The Romans added the pantheon of Greek gods to their own set of household gods, the *penates*, and the guardian spirits, the *lares*, only giving them new Latin names. The Roman gods had a human form and behaved like human beings, just like the Greek ones.[4] Roman mythology was presented in Virgil's *Aeneid*, Ovid's *Metamorphoses* and *Book of Days*, and Dionysius of Halicarnassus's *Roman Antiquities*.

After saving the Roman Republic from ruin in 48 B.C.E., Julius Caesar was posthumously granted the title of Julius the god by the Roman Senate.[5] His adopted son and heir, Augustus, became the first Roman emperor in 27 B.C.E. and called himself the son of a god.[6] Later emperors, such as Caligula and Nero, called themselves a living god and embraced being, or demanded to be, worshiped.[7,8]

THE GOD OF THE JEWS

Around 4,000 years ago, a man named Abraham obeyed the command of God to leave his city of Ur (in modern southern Iraq) for Canaan, the stretch of land along the eastern coast of the Mediterranean between the Nile River in Egypt and the Euphrates River in Iraq. This began the history of the Jews as a people and of their religion, called Judaism.[9]

What sets Judaism apart from the religions of nearby peoples is the direct intervention of a supreme god in the affairs of a specific chosen people, the Jews. This supreme god (Elohim, Yahweh, or Jehovah) is considered a real, living deity, expressing himself through his relationship with Jewish leaders and prophets, with whom he made covenants, or agreements. Table 4.1 gives a description of the covenants made by God with Abraham, Moses, and David.[10]

Table 4-1 Covenants between God and the Jews – Jewish Bible

Source in the Bible	God's Promise	Jews' Obligation	Sign of the Covenant
Genesis 9:8–17	Never again to destroy the earth and living creatures through flood	Avoiding idolatry, adultery, stealing, cursing God, and murder; avoiding eating live animal flesh; and establishing courts of justice	Rainbow
Genesis 12:1–7	To make Abraham the father of a great nation; all nations will be blessed through him	Loyalty to and faith in God	Male circumcision
Genesis 13:14–18	To give him and his descendants Canaan		
Genesis 17:1–7	To make him father of many nations; to make an everlasting covenant with the Jewish people		

Source in the Bible	God's Promise	Jews' Obligation	Sign of the Covenant
Exodus 19:5	Israel will be God's prized possession		
Exodus 19:6	Israel will be a kingdom of priests to God	Keep and follow the covenant	The sabbath
Exodus 19:6	Israel will be a holy nation		
Exodus 23:22	God will fight for Israel and vanquish all enemies		
Exodus 34:6–7	God will treat Israel with grace, mercy, and forgiveness		
2 Samuel 7	To establish David and his descendants as kings of Israel	David's faithfulness	Dynasty

Abraham, his son Isaac, and grandson Jacob were termed the patriarchs. Together with Moses, they are the main characters in the Torah (Pentateuch), the first five books of the Jewish Bible. The Torah is the sacred scripture of the Jews, just as the Old and New Testaments are to Christians. Moses is traditionally considered the author of the Torah, but most modern scholars find evidence for several authors. As a whole, the Torah tells of God's abiding role in the history of his beloved people.

THE JEWS

The Jews trace their origins from the many Semitic peoples living in the Mesopotamian region long ago. According to the *Cambridge Dictionary*, Semitic refers "to the race or people that includes Arabs and Jews, or to their languages." Like other Semitic peoples, the Jews (then called Israelites) had also worshiped many gods until Abraham chose the one, true god from among many.

ARCHEOLOGICAL FINDINGS

In 1991, the author and historian Martin Gilbert reported that there had been no archeological evidence that Abraham ever existed, nor any historical document showing that the Exodus of the Israelites from Ancient Egypt ever happened.[11]

In October 1999, the archeologist Ze'ev Herzog corroborated this, adding that the Israelites likely never conquered Canaan for redistribution among the twelve tribes of Israel, nor were had they ever been in Egypt as slaves. Further, Egypt ruled Canaan at the time of the supposed exodus. He described the gap between the archeological findings and the Biblical tales to be unbridgeable. Nor has any evidence of the existence of a King David or his heir Solomon been found.

The use of the technologies of carbon dating and other molecular technologies has led to no change: The Old Testament stories have no historical basis.[12]

5

Apotheosis II

"The Word became flesh to make us 'partakers of the divine nature': 'For this is why the Word became man, and the Son of God became the Son of man: so that man, by entering into communion with the Word and thus receiving divine sonship might become a Son of God.' 'For the Son of God became man so that we might become God.' 'The only-begotten Son of God, wanting to make us sharers in his divinity, assumed our nature, so that he, made man, might make men gods.'" [1]

— CATECHISM OF THE CATHOLIC CHURCH —

"I do not accept the orthodox teaching that Jesus was or is God incarnate in the accepted sense or that he was or is the only Son of God." [2]

— MOHANDAS K. GANDHI —

JESUS OF NAZARETH

Jesus was the founder of Christianity. He was born around 4 B.C.E. to a female virgin named Mary; at thirty years old, he began his public ministry with the help of twelve chosen disciples, preaching, performing miracles, and teaching. Because he offended the Jewish religious or civil authorities, calling himself the "Messiah, the Son of God" (John 19:7), he was charged with the capital offense of blasphemy and referred to the Roman procurator, Pontius Pilate, to be sentenced to die (for the crime of insurrection) by crucifixion. At thirty-three years old, Jesus died on the cross. However, after three days in the tomb, Jesus' body went missing: An angel told the three women who visited it that Jesus had been resurrected. Jesus then reappeared to his disciples and dwelt with them for some time before finally ascending to Heaven.

As before in the Bible, the supreme god intervened in the affairs of the Jews, but this time, he acted through his son, Jesus, a god-made man, come to live among his people. Jesus was a Jew. His genealogy can be traced to the house of King David, even to Abraham and further back to the first man, Adam.[3]

Jesus appealed to the Jews, but they rejected him, expecting a political leader to be the Messiah, who would liberate them from their oppressors and establish a righteous kingdom on Earth. Jesus was not this person that they were expecting. Additionally, Jesus upset the status quo and challenged the powerful, expressing sharp criticism of their high-handed conduct.

Jesus said: "Do not think that I have come to abolish the Law or the Prophets; I have not come to abolish them but to fulfill them" (Matthew 5:17). The Jews, however, had only one supreme god, who was revealed in the Bible. They did not support what Christianity believes, that Jesus was the mediator between the supreme god and humanity, Jews and Gentiles (non-Jews) alike. Jesus was bringing a new covenant, and this one was not exclusive to the Jews. Table 5-1 summarizes the aspects of this new relationship with God.

The New Testament describes God's love for all humanity, expressed through the sacrifice of his only begotten son, Jesus Christ, on the cross, without which salvation from death (sin) is impossible.[4]

Table 5-1 Covenants between Jesus and Humanity – New Testament

New Testament Book	The Lord's Promise	Humans' Obligation	Sign of the Covenant
John 14:16–17	To give the Holy Spirit	Repent	Baptism
Romans 10:9–10	To offer salvation		Eucharist
Mark 1:15	To bestow the Kingdom of God	Believe in Jesus as Lord and Savior and follow his commandments	Lord's Prayer
John 3:16	To give eternal life		
James 1:12	To grant the crown of life		

AUTHENTICITY OF THE FOUR GOSPELS

The gospel of Mark was probably written first of the four gospels, likely based on oral tradition, around 70 C.E., forty years after Jesus's death. The gospels of Matthew and Luke were patterned after Mark, using additional material as well, around 90 C.E., and the gospel of John likely came last, written in 100 C.E. The earliest New Testament was produced about 200 C.E. It was only recognized by the Church to be sacred scripture in 393 C.E. The precise authorship of the gospels remains controversial, and dubious sources and forgeries abound. A recent study by the historian Richard Carrier even called them nothing but myths.[5]

PAUL'S IGNORANCE

Only seven out of the thirteen epistles attributed to Paul in the New Testament have been confirmed to be written by him. The earliest of these was written in 50 C.E., twenty years after Jesus's death.[6] Because they were written earlier than the gospels, Paul's letters seem to bear a heavy burden of proving whether Jesus even existed at all. Scholars point out that these epistles do not contain references to anything told of Jesus in the gospels.[7]

Because the gospels and epistles are naturally considered to be biased, scholars have sought third-party sources to prove that Jesus existed historically. The most commonly cited references are from the Pharisee historian Josephus, writing in 93 C.E., and the Roman historian Tacitus, writing in 120 C.E. However, these sources have been disputed, allegedly altered or with later insertions. Beyond these, there is little evidence that Jesus even existed.[8]

ALBERT SCHWEITZER

In 1906, Albert Schweitzer published *The Quest of the Historical Jesus*, in which he studied the works of the eighteenth- and nineteenth-century scholars who sought to prove the historical Jesus. He claimed that, trapped within their subjective preconception of how they wanted Jesus to be, their studies were inherently flawed. In Schweitzer's opinion, Jesus began his public ministry with its end point in mind: that the end times (and the Last Judgment) were near and the kingdom of heaven was at hand; this led to the urgency of his message. Having developed this thesis, Schweitzer concluded the following:

> The Jesus of Nazareth who came forward publicly as the Messiah, who preached the ethic of the kingdom of God, who founded the kingdom of heaven upon earth and died to give his work its final consecration never existed. He is a figure designed by rationalism, endowed with life by liberalism, and clothed by modern theology in a historical garb. This image has not been destroyed from outside; it has fallen to pieces...[9]

6

The Worship of Man

"Ah, yes, superstition: It would appear to be cowardice in [the] face of the supernatural." [1]
— THEOPHRASTUS —

"We discover in the gospels a groundwork of vulgar ignorance, of things impossible, of superstition, fanaticism and fabrication." [2]
— THOMAS JEFFERSON —

"Faith must have adequate evidence, else it is mere superstition." [3]
— ARCHIBALD ALEXANDER HODGE —

MONOPOLY OF FAITH

By 200 C.E., Christianity had seeped into the consciousness of the Roman Empire, as dedicated disciples like Paul preached the Good News to audiences who were, according to Greek philosopher Celsus, mainly "slaves, women and little children."[4] The turning point for Christianity came in 313 C.E., when under the Edict of Milan, Emperor Constantine allowed Christians freedom of worship and restored their confiscated property. It was not long before they were building churches. Later Roman emperors after Constantine increased their support of Christianity, and in 380 C.E., Theodosius I took Christianity under state protection.[5]

Through the evangelism of intrepid friars—that is, the Dominicans and Franciscans, the appointment of bishops and abbots by the clergy in their respective dioceses or monasteries, the conversion of devout pagan rulers, and the development of important political alliances between kingdoms, Christianity grew rapidly. By 1200 C.E., it was the dominant religion in both Western and Eastern Europe, having a virtual monopoly on faith.[6]

SUBMISSION TO AUTHORITY

Jesus said he would found the Church through his disciple Peter, who was later understood to have been the first pope: "And I tell you that you are Peter, and on this rock I will build my church, and the gates of Hades will not overcome it.[19] I will give you the keys of the kingdom of heaven" (Matthew 16:18–19). The authority implicitly given to the Church by God himself could not be questioned. Its teachings must be followed. It brooked no opposition.

People believed that the social hierarchy was divinely ordained by God. One could do nothing to change one's station in life.[7] The Church even involved itself in daily life through the sacraments, from birth through marriage, death, and the afterlife. Each peasant and each serf was obliged to pay 10 percent of his gross income every year, in the form of either money or agricultural goods, to fund these clerical services, festivals, and more.[8]

Baptisms were performed in a baptismal font. In addition to this function, it was also used for the ordeal by water, which proved a person guilty or innocent. The accused was bound hand and foot and thrown into the font, and if he floated, he was innocent; if he drowned, he was guilty. Everyone was expected to be baptized and to leave a bequest to the Church at death; without this, one could not have a decent burial in the churchyard.[9]

The so-called ordeal by iron, in which a hot poker was given to the accused to hold, was also considered to be a foolproof test of guilt.[10]

The afterlife was presented as being made up of Heaven, Purgatory, and Hell. Dante Alighieri's poetic masterpiece, *The Divine Comedy*, presents allegorical visions of the world beyond the grave showing purgatory as a halfway house for souls trapped in limbo until their sins were repaid. To shorten their present or future agony, ecclesiastical papers, called indulgences, were sold at substantial sums of money by the clergy to the faithful for the remission (or absolution) of sins.[11] The sale of relics—saintly tokens like the hair or nails of a holy person or of holy water, statues, or amulets—was also a going concern; people looked to the saints for their protection.[12]

Every Sunday at least, priests celebrated Mass or the Eucharist in Latin, a language that the peasants did not understand, turning their back on them as they did so.[13] During consecration, the bread and wine literally turned into the body and blood of Jesus who had been sacrificed, in a process called transubstantiation.[14]

Ordinary life in the Middle Ages could be described as nothing but Hobbesian: "solitary, poor, nasty, brutish, and short."[15] Table 6-1 displays some medieval vital statistics.

The Church grew very wealthy because everyone who could, including emperors, kings, landlords, and nobles, donated vast treasures and landholdings to it to ensure their salvation and entry into Heaven. Monasteries were often the beneficiaries of the monks' family endowments or of lords' forfeiture of lands after death in war or for the nonpayment of loans. Because all its lands were inalienable, the Church became the largest landowner in Europe.[16] In addition, all tithes taken from the gross produce of secular land or from incomes were tax free.[17]

IN RETROSPECT

Christianity permeated the consciousness of Western Europe for 1,000 years. The Church monopolized education, producing learned clerks throughout the continent.[18] What little education there was revolved around the Bible, and scholarship focused on theology, to the exclusion of Greek science and philosophy.[19] Historian Will Durant summarized it in this way: "Having displaced the axis of man's concern from this world to the next, Christianity offered supernatural explanations for historical events, and thereby passively discouraged the investigation of natural causes; many of the advances made by Greek science through seven centuries were sacrificed to the cosmology and biology of Genesis."[20]

On the Byzantine Empire, which was relatively well insulated from the internal strife and external struggles ravaging its Western counterpart and lasted another 1,000 years to 1453 C.E., Durant had this to say: "There was something shallow about it (Byzantine civilization), a veneer of aristocratic refinement covering a mass of popular superstition, fanaticism, and literate ignorance; and half the culture was devoted to perpetuating that ignorance. No science, no philosophy, was allowed to develop in conflict with that ignorance; and for a thousand years no addition was made by a Greek civilization to man's knowledge of the world."[21]

Table 6-1 Vital Statistics – the Middle Ages (500 C.E.–1500 C.E.)

Middle Ages – 500 C.E. – 1500 C.E.[22]	
Life Expectancy	32
Child Mortality	30%
Literacy Rate	5%
Poverty Rate	95%

7

The Nature of Man

"Man is the measure of all things." [1]

— PROTAGORAS —

"We honor the Greeks because in their art, literature, philosophy and civic history we discern the early stirrings of our own ideals—rationalism, humanism, democracy— which first took firm root in Athenian soil." [2]

— CAROLINE ALEXANDER —

"A humanist has four leading characteristics—curiosity, a free mind, belief in good taste, and belief in the human race." [3]

— E. M. FORSTER

THE CHURCH IN CRISIS

Just as Constantinople, the capital of the Byzantine Empire, was falling to the Ottoman Turks in 1453 C.E., the Renaissance was coming to full flower in the city states of Florence, Milan, and Venice. This period, literally called the rebirth, saw a revival of interest in the history, philosophy, arts, and literature of ancient Greece and Rome, upon whose past glory the civilization of Western Europe had originally been founded. Even Rome took part in this movement; the popes fully embraced the invigorating and ennobling spirit it infused into cultural life. Further impetus was given by the new-found wealth from banking, commerce, and trade that was dispensed by beneficent patrons like the Medici family and the Church, and the fall of Constantinople brought intellectual Greek-speaking refugees and scholars trickling into southern Italy as well, bearing with them additional classical manuscripts from antiquity.[4]

However, just as Rome was at its peak, enjoying the artistic achievements of Leonardo, Raphael, and Michelangelo, the German scholar and monk Martin Luther was igniting a conflagration, which would come to be called the Reformation; by 1517 C.E., it would create a second schism in the Catholic Church, following the one that broke the Catholic from the Orthodox Church 463 years earlier, in 1054 C.E. The immediate reason for Luther's revolt was the sale of indulgences, which was being heavily promoted by Pope Leo X to defray the cost of building St. Peter's Basilica. Luther denounced this expedient in his "95 Theses," nailed to the door of the church in Wittenberg.[5] The main culprit was understood to be the breakdown of morals in the clerical ranks, including in the Roman curia, as they disavowed poverty, chastity, and obedience in their pursuit of worldliness—all of which was a reflection of the pope's poor leadership.

Compounding the division within, the Church was wracked by division from without. The emerging Catholic nation-kingdoms were restive: Each sought to wrest control of its churches and church lands away from grasping Rome, and they challenged the collection of tithes and taxes from their lands that were sent as tribute to Rome.[6] King

Francis I became the independent head of the Church in France in 1516 C.E. by taking advantage of Pope Leo X, and Spain had gained control over its churches in 1478 C.E., under the monarchs King Ferdinand and Queen Isabella. The hitherto tenuous union between Church and State in Germany, England, and Switzerland, which were rising economic powers in Western Europe, began to fray. One by one, the European kingdoms, pushed or led by reformers and rulers, disengaged from the orbit of Rome, culminating in the Peace of Westphalia in 1648.

HUMANISM

The consciousness of the latter Middle Ages swung quickly from the religious to the secular, from the supernatural to the natural, and from God back to man. Ideals of truth, beauty, and wisdom could be found in actual life amidst man's achievements at the same time as the Church was preaching that they could only be found in divine revelation. This sparked a widespread interest in revisiting classical antiquity as a rich repository of human knowledge that could teach, be imitated, and edify.

Considered a father of the Renaissance and perhaps the first humanist, the Italian poet and scholar Petrarch rediscovered a collection of Cicero's letters in 1345 C.E., whose very rediscovery sparked the fourteenth--century Renaissance; he made a copy of these letters and announced their discovery to the world by writing replies addressed to Cicero. His epic poem *Africa*, written in Latin, famously adhered to classical Latin style. He collected whatever old manuscripts he could find and even commissioned the first Latin translation of Homer. However, his famed love poems were written in the Italian vernacular.[7]

Among the most notable patrons of the Renaissance was the Medici family of Florence. Cosimo, the patriarch of the family, collected manuscripts; he would buy them if possible, but if not, he would have them copied, storing the copies in a monastery, an abbey, or his own library, all of which were open to the public. He engaged the services of the sculptor Donatello and founded what he called a Platonic Academy.[8] His grandson Lorenzo continued his legacy of collecting classical pieces; by 1495 C.E., altogether the collection included 1,039 volumes, of which 579 were Roman.[9] As the reputation of the Medicis for patronage grew, scholars began to come to Florence, studying and teaching at the University of Pisa, which Lorenzo had rebuilt.

Belief in the old science that was accepted by the Catholic Church became discredited. For instance, in 1543 C.E., Flemish physician Andreas Vesalius published his groundbreaking work *On the Fabric of the Human Body*, an accurate description of human organs and their functions, based on the dissection of actual human bodies, disproving the conjectures that Galen, a Greek physician, had proposed some 1,300 years previously based on animal dissections alone.

In the same year, the Polish astronomer and mathematician Nicolaus Copernicus published *On the Revolutions of the Celestial Spheres*, which proposed that the earth and all other planets revolved around the sun, in complete opposition to the accepted belief that the earth was at the center of the universe in a picture of the cosmos put forward by Ptolemy some 1,400 years earlier, in harmony with Aristotelian philosophy. In 1632 C.E., the mathematician Galileo Galilei confirmed Copernicus's theory but was condemned by the Church, forced to recant his heretical opinion, and sentenced to house arrest for the last nine years of his life.[10]

Man's ability to cater to necessity was expressed in the form of inventions. Manual copying of manuscripts on parchment gradually became replaced by printing with lead type on paper in the wooden printing press invented by the German goldsmith Johannes Gutenberg in 1440 C.E. This technology greatly aided the cause of the Reformation: Luther's tracts were quickly printed in large quantities, reaching a wide audience all at the same time. Luther also translated the Greek Bible into vernacular German, allowing the public to read what had once been the exclusive property of the clergy. This also helped break the stranglehold that the few had on education, developing literacy within the expanding middle class.

FRANCIS BACON

One of the greatest humanists of all arose in Renaissance England in 1561, under the name Francis Bacon. He was an accomplished lawyer, statesman, and philosopher. Although he was not a scientist, he expounded the philosophy of science. He had a vision, in his words, "to try the whole thing anew upon a better plan, and to commence a total reconstruction of sciences, (practical) arts, and all human knowledge raised upon the proper foundation."[11]

In his *Advancement of Learning*, he classified the desirable sciences and their particular problems, as well as reporting the research directions each required. Next, he proposed to bring Aristotelian logic up to date to perfect the use of human reason in his *New Organon*. He advocated the use of induction (in place of deduction), reasoning, observation, and experimentation in the study of nature. What he proposed effectively became the formula for the modern scientific method.

Although he was not able to see his grand project to completion, he inspired both future English philosophers, like Thomas Hobbes, John Locke, John Stuart Mill, and Herbert Spencer, and future scientists, who banded together through the Royal Society (1660 C.E.), following up his idea of scientists and researchers working together communally. Likewise, when the *Encyclopedia or a Systematic Dictionary of the Sciences, Arts and Crafts* was published in 1751 C.E., the French editors dedicated it to Bacon. In his immortal words, he is still ringing the bell for future development: "If any human being earnestly desire to push on to new discoveries instead of just retaining and using the old; to win victories over Nature as a worker rather than over hostile critics as a disputant; to attain, in fact, clear and demonstrative knowledge instead of attractive and probable theory; we invite him as a true son of Science to join our ranks."[12]

Image 7-1 Renaissance Man: Leonardo da Vinci (1452–1519). Italian painter, sculptor, engineer, astronomer, anatomist, biologist, geologist, physicist, architect, musician, philosopher, and humanist.[13]

8

Eureka!

"*Neglect of mathematics works injury to all knowledge, since he who is ignorant of it cannot know the other sciences or things of this world. And what is worst, those who are thus ignorant are unable to perceive their own ignorance, and so do not seek a remedy.*" [1]

— ROGER BACON —

"*Mathematics is not just a language. Mathematics is language plus reasoning.*" [2]

— RICHARD FEYNMAN —

THE CHURCH IN TRANSITION

In response to the challenge of the Reformation, the Church convened the Council of Trent (1545–1563 C.E.). It condemned Protestantism on all points, affirmed Catholic doctrines, initiated reforms to instill training and discipline in the clergy and to develop better administration, and upheld good works as necessary for salvation.[3] Reinvigorated, from 1566 to 1644 C.E., the papal conclave elected a succession of strong, judicious popes to see the reforms through.[4] It helped that Spain, a bulwark of the Catholic faith, entirely repudiated both the Renaissance and Reformation that preoccupied its European neighbors; the Spanish Inquisition (1478–1834 C.E.) enforced orthodoxy, and the Jesuit order brought fresh young troops to the embattled faith.

Founded by the Basque Ignatius of Loyola in 1534 C.E., the Society of Jesus (Jesuits) was one of the new religious orders that answered the call of the Church. The group was successful in outflanking the Protestants: Thanks to its work, Bohemia and Austria returned to the Catholic fold. Education was the specialty of the order, which, by 1640 C.E., had founded more than 500 schools, as well as some seminaries and universities.[5] "Their aim was apparently to produce an educated but conservative elite capable of intelligent and practical leadership, yet untroubled by doctrinal doubts and immovably rooted in the Catholic creed."[6]

In 1493 C.E., the Spanish Pope Alexander VI issued a papal bull that divided the world between Portugal and Spain through an imaginary north–south line placed west of the Azores in the Atlantic; all undiscovered lands west of the line belonged to Spain, and those east of the line belonged to Portugal.[7] Exploration of the New World soon began in earnest. Dominican, Augustinian, Franciscan, and Jesuit missionaries soon followed in the conquistadors' footsteps to spread the Christian faith among pagans in North and South America, Africa, and Asia.

By 1648, however, the Church had lost Germany, Scandinavia, Switzerland, the Netherlands, England, and Scotland to the Protestants. However, it had also gained millions of pagan converts from the distant lands outside Europe.

MATHEMATICS

Mathematics is the study of numbers, shapes, and other abstract patterns.[8] Where mathematical truths are expressed in equations, they incorporate problems in need of a solution. When they are solved correctly, the answers are impeccably precise, certain, agreeable, objective, clean, and reliable. This is the beauty of numbers. As a tool used in the study of nature, mathematics ultimately became indispensable for deciphering natural laws; once the order in nature is revealed, man can tame her to his advantage. Paraphrasing Galileo, to comprehend the secret of the universe, science understood that the words in the book were numbers, and their language was mathematics.

SCIENTIFIC REVOLUTION

In his book *On the Motion of Heart and Blood*, published in 1628 C.E., the English doctor William Harvey showed, through a description of a series of experiments on animals that measured the volume of their blood flow—which seemed to be too much to be continuously made in the flesh, as asserted in ancient medicine—that the blood pumped by the heart circulates all over the body and that the arteries carry blood away from the heart while the veins carry blood toward it.[9] Like Vesalius before him, Harvey's conclusions went against conventional wisdom, which was still based on Galen's practice. Although his conclusions were derived from inference (in animal testing and the anatomy of a human arm), his method had a sound scientific basis.[10]

Astronomy is the science that deals with the heavenly bodies, and physics concerns the nonliving natural world, including the heavens. For as far back as we know, from the Babylonians in 1,000 B.C.E., people have believed that the earth is at the center of the universe, in what is called the geocentric model, until Nicolaus Copernicus in the Renaissance period produced his account of the sun at the center of the world, in a heliocentric model. After the invention of the telescope, the later astronomers Galileo Galilei and the German Johannes Kepler confirmed his theory. Kepler's book *New Astronomy*, published in 1609, presented two laws: One described how the planets move around the sun in elliptical orbits, and the other accounted for their various speeds, faster when near it than away from it.[11]

The most celebrated scientific genius of the seventeenth century was the English physicist Isaac Newton. In his greatest accomplishment, he integrated the work of Copernicus, Galileo, and Kepler into a new depiction of how the universe operated in his book *Mathematical Principles of Natural Philosophy*, published in 1687 C.E. He proposed three laws of motion, which were grounded in Galileo's work, in addition to a law of gravity that explained Kepler's account of why the celestial bodies moved in the way that they did.[12] These laws apply not only to earthly phenomena but to everywhere in the universe. Laying the foundation for hundreds of years of physics, his work is some of the most profound of all time. Science would never be the same.

Table 8-1 Significant Inventions and Scientific Discoveries of the Seventeenth Century

Scientist	Inventions	Discoveries
Galileo Galilei 1564–1642	Hydrostatic balance	Four of Jupiter's moons
	Galileo's pump	Craters and mountains on the Moon
	Pendulum clock	
	Geometric compass	Stars in the Milky Way
	Thermometer	Phases of Venus
	Galilean telescopes	Sunspots
		The principle of pendular motion
		Law of falling bodies
		Principle of inertia
		Relative concepts of acceleration, force, and mass

Scientist	Inventions	Discoveries
Johannes Kepler 1571–1630	Keplerian telescope Eyeglasses The logarithm	The three laws of planetary motion The moon as the cause of the tides in the ocean
Isaac Newton 1642–1725	Reflecting telescope Calculus The milled coin	The three laws of motion The law of universal gravitation Understanding of celestial mechanics Sunlight as made up of a light spectrum Light as a particle

9

Sapere Aude!

"He who will not reason is a bigot; he who cannot is a fool; and he who dares not is a slave." [1]

— WILLIAM DRUMMOND —

"The apple cannot be stuck back on the Tree of Knowledge; once we begin to see, we are doomed and challenged to seek the strength to see more, not less." [2]

— ARTHUR MILLER —

THE CHURCH IN DECLINE

The Reformation culminated in the Thirty Years War (1618–1648 C.E.). It began as a local conflict between Catholic and Protestant partisans within the Kingdom of Bohemia; soon the theater of war grew to engulf the greater part of Europe, as the political alliance of Spain and the Holy Roman Empire was pitted against the loose coalition of Denmark, Sweden, France, and the Protestant German principalities. By the time the Peace of Westphalia was signed in 1648 C.E., an estimated one-third of the Holy Roman Empire's population was dead.[3] Ultimately, it was a Protestant victory. Spain became bankrupt, and Rome became politically powerless.

On the other hand, England's own Civil War raged from 1642–1651 C.E. under the reign of King Charles I; the armies of the monarch and Parliament engaged in pitched battles to determine whose right to rule was supreme. The Parliamentarians won, installing Oliver Cromwell as the head of a republican government. Around 10 percent of Britain's population was lost.[4]

Meanwhile, England and the Netherlands maintained and expanded their overseas presence, which had been spearheaded by commercial trading in Africa, India, and the East Indies in the seventeenth century and later, slowly, through colonies in North America.[5]

At the turn of the eighteenth century, Central Europe was still exhausted from the Thirty Years War. People became greatly disillusioned with Christianity.

> But though the Reformation had been saved, it suffered, along with Catholicism, from a skepticism encouraged by the coarseness of religious polemics, the brutality of the war, and the cruelties of belief … Men began to doubt creeds that preached Christ and practiced wholesale fratricide. They discovered the political and economic motives that hid under religious formulas, and they suspected their rulers of having no real faith but the lust for power … Even in this darkest of modern ages an increasing number of men turned to science and philosophy for answers less incarnadined than those which the faiths had so violently sought to enforce … The Peace of

Westphalia ended the reign of theology over the European mind, and left the road obstructed but passable for the tentatives of reason.[6]

THE ENLIGHTENMENT

The Enlightenment was the name given to the following period of rational reevaluation of established institutions (including monarchy and the Church), politics, and society, in light of what had transpired over the previous century. Intellectuals, in particular, questioned the status quo: its beliefs, pretentions, and authority. Following Newton's description of a clockwork universe, later thinkers embarked on a more scientific approach to philosophy, politics, religion, and many other subjects that piqued their fancy.

Foremost among these thinkers were the Englishman John Locke and the Scot David Hume. Locke's writings centered on politics and knowledge, while Hume investigated knowledge and human nature, a pursuit that emerged as skepticism. Another Scot, Adam Smith, laid the foundations of economics as a science.

On the opposite side of the English Channel, three Frenchmen stood out in this period: Jean-Jacques Rousseau, Charles Montesquieu, and Voltaire. Rousseau and Montesquieu targeted the French monarchy, the former advocating a new social contract between the ruler and the governed; the latter called for a separation of powers among the three governing branches of state. Voltaire was the *enfant terrible* of the powers that be, the personification of the anti-*ancien regime*. His satirical commentary can be seen in his novel *Candide*, published in 1759, along with about 200 books and thousands of letters.

Finally, the Dutch Jew Baruch Spinoza and the German Immanuel Kant considered the relationship between religion and morality. Both believed in the existence of an impersonal God. To comprehend God and reach the highest virtue, Spinoza promoted the use of reason to enable a blessed, happy life. For Kant, even without religion, morality, guided by practical reason and conscience, could lead to virtue.

INDIVIDUALISM

As the secular thinking of the Enlightenment found its way to the broader public through books, treatises, and tracts, people began to think and question for themselves to a greater degree. The Bible translations into the vernaculars aided this, as people began to be able to read the Word of God for themselves in the comfort of their homes. A trend of turning away from the ritualistic forms of church worship and toward individual self-reflection began to sweep Europe. Many Protestant groups developed evangelical revival movements, with ministers preaching, teaching, converting, and helping the poor. The grassroots movement of Pietism sprouted in Germany, and in England, the similar movement of Methodism thrived.[7]

Across the Atlantic, the First Great Awakening swept through American colonies, led by charismatic ministers; Presbyterians established a foothold in Virginia, and the Baptists spread across the Carolinas. Soon, missionaries came to the Native American Indians. Elsewhere, such Protestant movements led to the establishment of Bible colleges and universities.[8]

Table 9-1 Eighteenth-Century Philosophers and Their Key Ideas

Philosopher	Study	Key Ideas	Influenced
John Locke 1632–1704	Philosophy Politics	Consciousness Knowledge through experience alone Religious tolerance All humans created equal Right to defend life, liberty, and property Separation of the powers of the state Separation of church and state Right to revolt	US Declaration of Independence Liberalism The republican state Political philosophy American Revolution

Philosopher	Study	Key Ideas	Influenced
David Hume 1711–1776	Philosophy History	Knowledge through experience alone Self as a bundle of experiences Passion, not reason, governs behavior Ethics as based on feelings Natural world without a designer Miracles do not happen Causal inferences based on custom or habit Freedom compatible with necessity	Western philosophy
Adam Smith 1723–1790	Economics Ethics	Self-interest + competition = prosperity The laissez-faire market Division of labor Business interest vs. public interest Sympathy as source for morality	Economics as a science Capitalism
Jean-Jacques Rousseau 1712–1778	Philosophy Politics Education	Civilization as inimical to man's progress Pride causes insecurity in society Morality vs. duty in the civic state Social contract between a government and its people General will vs. individual will Teaching reasoning as education Religious tolerance and deism	French Revolution American Revolution The republican state

Philosopher	Study	Key Ideas	Influenced
Charles Montesquieu 1689–1755	Politics	Separation of state powers Classification of political systems	US Constitution The republican state
Voltaire 1694–1778	Politics Religion Literature	Religious toleration Satirical commentaries on the status quo Freedom of speech Enlightened government Deism by reason	Western philosophy and literature
Baruch Spinoza 1632–1677	Philosophy Ethics	God and nature are one substance Everything happens through necessity Humans have no free will Nothing happens by chance Reality is perfection Knowledge of God and nature is the highest virtue	Pantheism Rationalism
Immanuel Kant 1724–1804	Philosophy Ethics	Synthesis of rationalism and empiricism The mind as an active participant in knowledge Morality as derived from rationality The categorical imperative or duty Treat others as ends, not as means Practical vs. pure reason Appearance vs. reality	Western philosophy

10
Prometheus Unbound

"Real knowledge is to know the extent of one's ignorance." [1]

— CONFUCIUS —

"Who is more humble? The scientist who looks at the universe with an open mind and accepts whatever the universe has to teach us, or somebody who says everything in this book must be considered the literal truth and never mind the fallibility of all the human beings involved?" [2]

— CARL SAGAN —

THE CHURCH IN FREE FALL

It did not take long for the liberal ideas of the Enlightenment to bear real-world political fruit. In 1776 C.E., the American Revolution began with its Declaration of Independence, inspired by the ideas of John Locke: rephrasing his proposed natural rights of man to "life, liberty, and the pursuit of happiness" and alluding to his assertion of the right to revolution against an abusive government, also following his proposed separation of church and state, which is now enshrined in the Constitution of the United States. Most of those who were most important in the American Revolution are believed to have been Deists.[3]

Thirteen years later came the explosion of the French Revolution, influenced by Jean-Jacques Rousseau's idea of a social contract, which denied the divine right of the monarch to rule, devolving that sovereign right instead to the people. It also produced its own Declaration of the Rights of Man and the Citizen, which appeared in the French Constitution in 1791 C.E., holding that all "men are born and remain free and with equal rights," asserting that the duty of the government was to preserve these "natural and inviolate rights ... liberty, property, security, and resistance to oppression," embodying the ideals of both Rousseau and Locke.[4]

Under the new French Republic, the power of Catholicism, which had been the state religion, underwent severe curtailment. It was identified too strongly with the monarchy, corruption, and landed wealth. To begin with, church property was confiscated and clerical privileges abolished. Then, Catholicism was outlawed, and the 30,000 Catholic clergy were exiled, jailed, or killed.[5] After the failure of a secular ideology put in place to replace Christianity, the government reinstituted Catholicism but formalized the separation of church and state in 1795 C.E.

The Church enjoyed a period of relative liberty before Napoleon Bonaparte came to power in 1799 C.E. He subordinated the church to the state in his concordat with the papacy in 1801 C.E. He later occupied Rome in 1808 C.E. and had Pope Pius VII imprisoned until his exile in 1814 C.E.[6] Pius VII then returned to Rome triumphant, bearing a certain halo of vindication in the eyes of the faithful after the years of suffering, and the Catholic Church thereafter experienced a revival, not only in France but in the rest of Catholic Europe as well.[7]

In the First Vatican Council in 1870, Pope Pius IX declared the doctrine of papal infallibility in faith and morals—that is, the Holy Spirit was asserted to guide the pope to keep him from falling into error—and papal supremacy, that is, that the pope was not subject to any secular authority in his papal decrees, nor was he accountable to any ecclesiastical authority other than his own. He also promulgated the dogma of the Immaculate Conception of Mary, the Mother of God.[8]

By the turn of the twentieth century, the entire world had been explored and colonized, the continents mapped, and the oceans navigated. Christianity spread far and wide, as wave upon wave of enthusiastic missionaries saw the vast opportunity to evangelize virgin territories and reach peoples who had not yet heard the Good News.[9]

SCIENCE AND CREATION

The term "scientist" was first attested in 1833 C.E. From that point on, science was understood to be a serious profession, and it featured increasing specialization in a range of fields of interest, whereas it had previously been considered to be the hobby of a small intellectual elite.

Biblical scholars have calculated the age of the earth from scripture at around 6,000 years old, beginning with its creation.[10] Geologists sought answers from the fossils found in rock layers, and the English geologist Charles Lyell concluded that the planet was even older than that. In 1862 C.E., the physicist Lord Kelvin published calculations estimating it to be between 20 million and 400 million years old.[11] The latest methods of radiometric dating put the origin of the earth at 4.5 billion years ago.

The Biblical creation story implies that animals, plants, and humans are all now just as they were created at the beginning. The naturalist Charles Darwin opposed that view, making a scientific case against it in his groundbreaking work *On the Origin of Species by Means of Natural Selection*, published in 1859. His theory of evolution contended that, in their struggle for existence, organisms evolved abilities, qualities, and advantages honed and tested through millions of years for them to adapt to their ever-changing and unforgiving environment; these same survival-oriented variations were passed on to their offspring, allowing them a better-than-average chance of flourishing.

His later work, *The Descent of Man*, published in 1871, asserted the origin of man in the apes.

Although they were contemporaries, Darwin was unaware of the Austrian friar Gregor Mendel, who conducted experiments with pea plants from 1856–1863 C.E. to establish the inheritance of physical attributes. He asserted two laws: the Law of Segregation and the Law of Independent Assortment, but his publication of the results in 1866 was ignored. He said that traits were inherited by means of invisible factors, which we now call genes.[12] Darwinism had no supporting mechanism for how heredity functioned until Mendel's work was rediscovered in 1900 C.E.

In the Bible, the story of Noah's Ark appears, in which Noah gathers all animals and birds, a pair of each kind, into his ark (Genesis 6:19–22). Currently, on Earth, there are about 1.5 million cataloged species, with around seven million more uncatalogued.[13] Even if we exclude the marine animal species and plant species, it would have been quite a task for Noah to have fit samples of all species into his ark.[14]

INDUSTRIAL REVOLUTION

The shift from agricultural to industrial production produced a virtuous circle of unprecedented prosperity in the world economy. By the end of the eighteenth century, innovation and technological breakthroughs in agriculture made food surpluses possible. This increased productivity led to lower prices for farm goods to feed the growing population. The attendant surplus in farm labor was diverted to the needs of the industrial market.

Britain was the first nation to industrialize. At the same time, Adam Smith's idea of the free market and its invisible hand was made the ideal. The invention of the steam engine was more responsible for the beginnings of the Industrial Revolution than any other piece of equipment.[15] Smith's fellow Scot James Watt made this machine more powerful, efficient, and cost effective in 1775 C.E. In addition to pumping water, its first use, Watt's new and improved version could power locomotives, ships, and other large, powerful machines.

A confluence of factors gave Britain the advantage in industrial development: its of abundance of coal and iron; its network of rivers, waterways, and canals; the long coastlines available for trade; and its well-developed railways. Its knack for logistics made increased urbanization possible; factories relocated to cities, which in turn attracted migrant workers.[16] Supported by bank financing, the novel joint-stock corporations enabled entrepreneurs to develop businesses and reinvest their profits in them.

The first wave of industrialization centered on the textile industry, whose cheap raw materials were sourced both locally, like wool, and internationally, like cotton imported from the southern United States. The second wave was oriented toward higher value-added products, like chemicals, cement, engineering, and steel, made possible by the Bessemer process.

By 1850 C.E., Britain owned half the world's ocean-going ships.[17] Its strength as a naval power facilitated its sourcing of raw materials and the transport of finished goods, an advantage both in trading with the rest of the globe and within its vast colonial empire, the world's largest.

Following Britain's template, Germany, Switzerland, and the United States industrialized, building and expanding rail networks in 1840–1870 C.E. Russia, Sweden, and France followed later in the century.

Table 10-1 Scientific Discoveries and Historic Inventions from the First Century to the Twentieth Century

Century	Number of Scientific Discoveries[18]	Number of Historic Inventions[19]
1st	1	4
2nd	4	2
3rd	3	3
4th	3	7
5th	11	4
6th	2	5
7th	8	4
8th	0	1
9th	6	2
10th	4	2
11th	7	4
12th	3	1
13th	1	7
14th	8	4
15th	1	8
16th	17	5
17th	22	10
18th	17	33
19th	43	98
20th	71	79

11

Audeamus!

"All is mystery; but he is a slave who will not struggle to penetrate the dark veil." [1]

— BENJAMIN DISRAELI —

*"We shall not cease from exploration
And the end of all our exploring
Will be to arrive where we started
And know the place for the first time."* [2]

— T.S. ELIOT —

"Mystery creates wonder and wonder is the basis of man's desire to understand." [3]

— NEIL ARMSTRONG —

THE CHURCH IN REMISSION

The Enlightenment affected Christianity, prompting the development of a new direction in Christian thinking that was based on reason and pragmatism, called liberal theology. This approach emphasized the personal experience of the divine and the autonomy of the individual on matters of faith over that of the Church and, hence, prioritized the individual's right to reason above the authority of both the Church and the Bible. In addition, it rejected the supernatural, including the dogmas of the Incarnation, miracles, and the doctrine of the Trinity. Faith should focus on how man can achieve God-consciousness (as Jesus, as man, did), that is, on the here and now rather than the afterlife.[4] This thought did not consider itself complete but rather as a work in progress. It believed that Christianity should change with the times. At the turn of the twentieth century, liberal theology was at its most influential, before the orthodox Catholic and Protestant churches successfully defended their faiths from further modernist encroachment.

While Europe was reeling in the aftermath of World War I, new major ideologies were gaining currency and challenging Christianity in one form or another: Nazism, socialism, and communism. Pope Pius XI signed a concordat with Adolf Hitler in 1933 C.E., hoping to preserve the rights of the Church in Germany, but this was to no avail; Hitler persecuted Christians anyway.[5] During the Spanish Civil War (1936–1939 C.E.), the Church sided with the nationalist General Francisco Franco against the socialists. Soviet Communism advocated atheism and was therefore an existential threat to Christianity wherever communists sowed their seed of revolution. World War II laid waste to most of Europe. It laid a crushing burden on peoples' spirits and on their religious faith, with two world wars less than one generation apart. As Europe's economies rose from the ashes, the secularization of its society only intensified and even accelerated.

However, Christianity continued to thrive in the developing countries of Africa, Asia, and Latin America. Not only was the Church busily preaching the gospel, it was also actively engaged in humanitarian causes; working to alleviate poverty; demanding social justice; and educating, sheltering, and feeding the marginalized in society. On the other hand, the Protestants gained renewed strength, especially in the United States, in the evangelical movement, which emphasized Bible study, fundamental doctrines of the faith, missionary activities, relief efforts, schools for ministers, and faith-oriented publishing companies.[6]

Pope John XXIII convened the Second Vatican Council (1962–1965 C.E.) to modernize the approach of the Church to presenting its message and seeking out its Christian brothers for cooperation in bettering the world. As part of this move, the celebration of the Eucharist came to be conducted in the vernacular, not only in Latin, and laypeople took on expanded roles as educators, parish administrators, Eucharistic ministers, readers, and coordinators. Another reform addressed the composition of the Roman Curia and the College of Cardinals, changing it from a Eurocentric, even Italian-centric, body to having a more global face.[7]

THE ERA OF SCIENCE AND TECHNOLOGY

Enormous advances in science and technology characterized the twentieth century. New knowledge in the material sciences, engineering, and equipment, among other areas, combined to make the two world wars the deadliest in history. War is the ultimate national competition, and the best minds were harnessed to gain that extra edge in pursuit of final victory. Some World War I-era technology that found its way to commercial use afterward led to the trench coat, blood banks, canned food, mobile X-ray machines, sanitary pads, Kleenex, stainless steel, the zipper, the wristwatch, and the drone.[8]

Some World War II-era technology had the following later commercial use: night vision, duct tape, sunglasses, walkie-talkie, radar, jet engine, computer, ballistic missile, and nuclear technology.[9] Albert Einstein's well-known equation $E = mc^2$, first published in 1905 C.E., became part of the basis for the atomic bomb, completed in 1945 C.E.,[10] and the V2 technology of German rocket scientist Wernher Von Braun was later utilized by the NASA space programs that finally put the first man on the moon.[11] Penicillin was accidentally discovered by the Scottish biologist Alexander Fleming in 1928, but owing to logistical complications, it could only be produced on an industrial scale in 1944; however, it saved the lives of many injured Allied soldiers.[12]

The beginnings of the internet can be traced to the U.S. military-funded project ARPANET, developed for time-shared use of computers in the 1960s.[13] The English computer scientist Tim Berners-Lee invented the world wide web as a source of information services in 1990 C.E.

The American electrical engineer Jack Kilby invented the first prototype of the integrated circuit (IC) made from germanium in 1958 C.E. A newer, more practical IC came in 1959 C.E., made from silicon, according to the design of fellow American physicist Robert Noyce;[14] all future versions of the microchip would be based on Noyce's invention. The microchip is now found in all electronic devices, such as the computer, smartphone, air conditioner, microwave, and car.

COLD WAR

Soon after World War II ended in 1945, another war began, an ideological confrontation between democracy or capitalism and communism, often called the Cold War, with the US and the USSR as primary protagonists, and it overshadowed the world.[15] These countries were superpowers based on their giant stockpiles of nuclear weapons, a brainchild of science gone awry, and these weapons of mass destruction grew exponentially more powerful than the ones that had been used in Japan. The 1962 Cuban Missile Crisis, the Korean War, the Vietnam War, and the Afghanistan War were confrontation and proxy wars that these superpowers participated in or supported. The one-upmanship between the two was even seen in sports competitions, the space race, and the development of military weaponry. However, the Berlin Wall fell in 1989, and the USSR imploded soon thereafter.

NATIONALISM

Empires all over the world began to disintegrate in modern times, perhaps beginning with the Ottoman Empire in the late nineteenth century, with its Balkan Wars, found with its subordinate peoples who demanded independence; it disappeared completely after World War I, together with the Austro-Hungarian Empire and the Russian Empire.[16] Similarly, the aftermath of World War II saw the former colonies of the imperial powers gain the right to self-determination, beginning with India and Pakistan, in Southeast Asia and then to Africa, where the British, French, Dutch, Belgian, Portuguese, and Spanish had remnant colonial claims.[17] When Britain handed back Hong Kong to China in 1997, and Portugal gave up Macau in 1999, the last vestiges of imperialism finally vanished. It is possible to understand the real impetus for the two world wars to have been in nationalism.

SCIENCE MYSTERIES

Two mysteries that science has not yet solved are the origin of life and the origin of the universe.

In 1927, the Belgian priest Georges Lemaitre published his finding that the universe is expanding, using Einstein's general theory of relativity. Two years later, the American astronomer Edwin Hubble independently observed that the multitudes of galaxies in outer space were moving away from the earth, thanks to the evidence of the red shift in their electromagnetic spectra.[18]

In 1931, Lemaitre proposed that the universe had begun from a singular point, the Primeval Atom, an idea that soon became popularly known as the Big Bang.[19] According to this theory, the universe is 13.8 billion years old and began as a hot and dense amount of pure energy before it burst into an expansionary mode. As it grew, it cooled, and energy particles formed the first protons and neutrons, which combined to form the atomic nuclei of hydrogen and helium; over time, electrons combined with the nuclei to produce the first atoms.[20]

Astronomers currently believe that the visible universe is only 5 percent of the total whole; the remainder is composed of dark energy (68 percent) and dark matter (27 percent).[21] Dark energy is a sort of anti-gravity, which is pushing the galaxies away from Earth. Dark matter is invisible matter, which does not emit or reflect light, but whose gravity affects visible matter; however, it has not yet been directly detected.[22]

There are four fundamental forces of nature: gravity, the strong nuclear force, the weak nuclear force, and electromagnetism. The general theory of relativity explains gravity, and quantum theory explains the other three forces.[23] However, the two theories are incompatible. Because all forces were created immediately after the Big Bang, theoretical physicists are working to recombine these forces into one single force, hoping to create a Theory of Everything; doing so would prove how the universe began before the Big Bang and would explain the behavior of all matter after it. Meanwhile, a Grand Unified Theory seeks to combine the three forces, excluding gravity, into one single force.[24]

The Large Hadron Collider, built in 2008, is the largest particle accelerator in the world. Scientists are using it to recreate the conditions obtained immediately after the Big Bang by colliding beams of high-energy protons, electrons, or ions at nearly the speed of light.[25] These experiments may lead to further discoveries on the nature of energy and matter as they have been since the beginning of time; thus proving, disproving, or forcing a reformulation of existing theories of the origins of the universe.

The fossil evidence suggests that life began around 3.5 billion years ago.[26] However, scientists do not yet know how this happened.

One theory suggests that lightning might have sparked the formation of amino acids and sugars in a primeval atmosphere filled with methane, ammonia, hydrogen, and water.[27] Another theory suggests that early life might have evolved from deep-sea hydrothermal vents that are rich in mineral elements and spew hot water, which could have interacted with chemical molecules to create amino acids and sugars; at those depths, life would have been shielded from ultraviolet rays because of the absence of an ozone layer.[28] A third, somewhat outlandish, theory suggests that life came from outer space via comets or meteorites; this idea is called panspermia.[29]

At this point, no one really knows, and scientists do not even have a shared definition of life.[30]

Table 11-1 Selected Scientific Discoveries and Inventions of the Twentieth Century

Scientist	Study	Discoveries
Joseph John Thomson 1856–1940	Physics	The electron (subatomic particle)
Ernest Rutherford 1871–1937	Physics	The atomic nucleus and the proton (subatomic particles)
Albert Einstein 1879–1955	Physics	The Special Theory of Relativity and $E = mc2$
		The General Theory of Relativity
		The Law of Photoelectric Effect
Alexander Fleming 1881–1955	Biology	Penicillin (antibiotic)
Max Planck 1858–1947	Physics	Quantum theory
Georges Lemaitre 1894–1966	Astronomy	The Big Bang Theory
J. Robert Oppenheimer 1904–1967	Physics	The atomic bomb
Wernher Von Braun 1912–1977	Aeronautics	Apollo Saturn V Rocket
Francis Crick 1916–2004	Biology	The structure of DNA
James Watson 1928–		
Jack Kilby 1923–2005	Physics	The microchip or integrated circuit
		The handheld calculator and the thermal printer
Robert Noyce 1927–1990	Physics	The silicon-based integrated circuit
Tim Berners-Lee 1955–	Computer Science	The World Wide Web

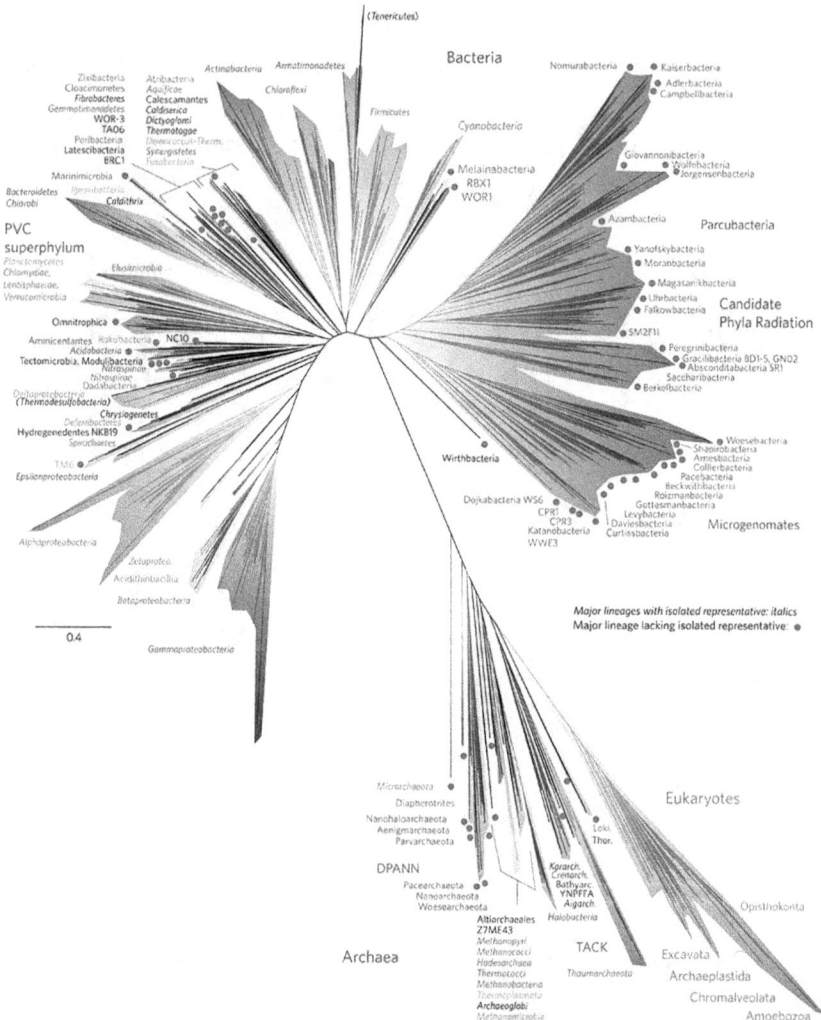

Image 11-1 Tree of Life based on the sequenced genome. The eukaryotes (multicellular organisms as plants and animals) are seen on the bottom right branch. The bacteria are the largest branch seen at the top.[31]

Image 11-2 Graphical representation of the Milky Way. Our solar system is a tiny speck of dust in the larger universe (indicated by the arrow) and lies near the peripheral edge of the galaxy's outer spiral.[32]

12

On Morality

"It seems to me that the idea of a personal God is an anthropological concept which I cannot take seriously. I also cannot imagine some will or goal outside the human sphere ... Science has been charged with undermining morality, but the charge is unjust. A man's ethical behavior should be based effectually on sympathy, education, and social ties and needs; no religious basis is necessary. Man would indeed be in a poor way if he had to be restrained by fear of punishment and hope of reward after death." [1]

— ALBERT EINSTEIN —

John Locke believed that without religion, the social order would break down. However, Napoleon cynically said that religion is "what prevents the poor from killing the rich,"[2] as was borne out by the bloody excesses of the French Revolution. Morality does not have to depend on religion, so it can stand on its own; the converse, however, is not true. Morality is not derivative of religion. In the first place, primitive religions, such as animism, began without reference to morals, because the idea of worship was to appease the spirits or gods of nature. Society can survive and order be maintained without religion through the active operation of the following:

Conscience. In general, man is intrinsically good. Cognitive ability is like a moral compass to people to be able to tell right from wrong, even before a moral judgment is rendered.

The Golden Rule. The maxim "Treat others as you want others to treat you" has been recognized by all cultures since antiquity.

Civil and criminal law. The earliest form of this was among the Ancient Babylonians in the Code of Hammurabi, dated to 1754 B.C.E., which consisted of 282 laws that should be followed for a just and orderly society. Then as now, laws are integral part of government.

Customs/mores. Social norms of what is unacceptable and acceptable behavior support discipline.

Sympathy. Adam Smith's *Theory of Moral Sentiments*, published in 1759, explored sympathy as a moral agent people use to investigate the mutual sympathy of sentiments, which forms the basis of conscience.

Education. Values and principles of virtue are taught in school and at home in contemporary society, the most literate set of human beings that the world has ever known.

The categorical imperative or moral duty. In his 1785 book *Groundwork of the Metaphysics of Morals*, Immanuel Kant wrote that man, who is rational and free, is bound by duty to always do what is right, following a universal law, as well as always treating others as an end, not as a means. Distinguishing morally right from morally wrong can only be done through the use of reason.[3]

Humanism. This is a secular movement dedicated to the greater good of humanity, which seeks to empower people to live ethically and with dignity.[4]

"Meanwhile, much of our moral freedom is good: it is pleasant to be relieved of theological terrors, to enjoy without qualm the pleasures that harm neither others nor ourselves, and to feel the tang of the open air upon our liberated flesh."[5]

13

Conclusion

"We have two lives, and the second begins when we realize we only have one." [1]

— CONFUCIUS —

*"**Napoleon,** when hearing about Laplace's latest book, said, 'M. Laplace, they tell me you have written this large book on the system of the universe, and have never even mentioned its creator.'*
***Laplace** responds, "Je n'avais pas besoin de cette hypothèse-là.' (I had no need of that hypothesis.)"* [2]

— ANECDOTE INVOLVING NAPOLEON AND FRENCH PHYSICIST PIERRE-SIMON LAPLACE CIRCA 1802, AS RECOUNTED BY MATHEMATICIAN JOSEPH LAGRANGE —

Based on the preceding historical survey of the development of Christianity in the Western hemisphere, we can draw some hypothetical conclusions, as follows:

1. All religions, including animism, mythology, polytheism, and monotheism, are human inventions created to try to make sense of nature and the fate of human beings, over neither of which humans have any control. "Apparently 'it was fear that first made the gods'—fear of hidden forces in the earth, rivers, oceans, trees, winds and sky. Religion became the propitiatory worship of these forces through offerings, sacrifice, incantation and prayer."[3]

2. Religion became part of shared culture around the time that people began to use modern speech or language about 68,000 B.C.E., allowing cooperation initially in larger bands, then tribes, villages, towns, and the ever larger polities of cities, city-states, and kingdoms, and then to nations and nation-states.

3. By 3,000 B.C.E., humans had made significant strides in managing their surroundings through their inventions, which benefited them sufficiently to convince them of their dominion over nature.

4. Human beings began projecting their ego outward onto a pantheon of gods, such as the Egyptian and Greek gods. These gods were created in the human image and with a human likeness.[4]

5. From the Jews, two religions came forth that were unlike anything seen until that point, involving living gods, not the ancient dead gods of other religions. Sometime around 2,000 B.C.E., Yahweh, God Almighty, revealed himself to Abraham; 2,000 years later, his incarnate son, the Lord Jesus Christ, came to Palestine and lived among the people there.

6. Jewish history was focused on the Promised Land or Palestine; this endemic nature means that the biblical narratives were meant only for the Jews.

7. The interactions of the Jewish God with human beings are recorded in the man-made texts of the Old and New Testament of the Bible,

which appear to have a questionable relationship with reality as understood by modern science and historical research.

8   The miracles as recounted in biblical tales are likely false. There are probably not many modern-day miracles, and in their recorded form, they appear to defy the laws of physics. These may have been deployed as ruses to win new impressionable converts into the religious fold.[5]

9   Christopher Hitchens, in the prologue, sensibly contending that one group would come into contact with an ultimate supernatural being not only once but twice, appears not to be credible.

10 Christianity, owing to theological disagreements over the Messiah, among other things, was a Jewish faction at first that broke from mainstream Judaism and taught a rival doctrine. This is not unlike the way the Byzantine Church broke away from the Catholic Church 1,000 years later (although this may be attributable more to cultural differences). Adolf Hitler stated: "The heaviest blow that ever struck humanity was the coming of Christianity ... (an) invention of the Jew."[6]

11 The Church grew to be the most powerful entity in Western Europe in the Middle Ages; it became the largest landowner in Europe and collected tithes from all.[7] "Though the Church served the state, it claimed to stand above all states, as morality should stand above power ... Claiming divine origin and spiritual hegemony, the Church offered itself as an international court to which all rulers were to be morally responsible."[8]

12 In the Middle Ages, 90 percent of the European population were serfs, and only 10 percent were freemen (landed peasants, merchants, artisans, soldiers, clergy, and nobles).[9] Most people were illiterate, even the nobility. The faithful were often instructed through the imagery of stained-glass windows, painted walls, and tiled reliefs of the cathedral, which depicted Bible scenes.[10] The people were highly impressionable and readily believed what they were told by those in power.[11]

13 The rediscovery of the glorious achievements of previous civilizations during the Renaissance (1434–1534 C.E.) impelled men to go further, first in the arts and then the sciences, which eventually led to the Scientific Revolution. More knowledge was acquired to better the lot of humankind in one hundred years than was achieved under the auspices of the Church in the previous 1,000. Galileo reflected that "I do not feel obliged to believe that the same God who has endowed us with sense, reason and intellect has intended us to forgo their use."[12]

14 Secular ideologies challenged and weakened the Church over the course of the eighteenth century, as a result of the Enlightenment, and during the twentieth century, owing to political ideologies. To this, the unrelenting practical influence of science on people's lives can be added, leading to a rapid loss of relevance for the Church.

15 Novel secular and scientific ideas will only grow stronger as the new knowledge discovered in one generation forms the foundation of the next and then the next in a virtuous cycle; this synergy is exponential and unstoppable.

16 The veil of the Catholic Church, "rich in miracle, mystery and myth,"[13] as exposed by the Reformation, is increasingly being peeled away by science. Its unchanging dogma and doctrine do not reveal anything new for the faithful; the same cast of characters continues from 2,000 years ago, as laid out in the Bible. In some ways, the Church has become anachronistic. It is hard to reinvent when there are few new concepts in a self-contained system. This may be why many ecumenical, born-again, revivalist movements have come and gone in Christianity.

17 The evangelization of the Church in less developed countries, targeting the poor and illiterate (as in the Middle Ages), is a worthy goal but acts to the detriment of pagan converts individually and of nations collectively; evangelization is spreading the Church's errors far and wide. "The poor you will always have with you, but you will not always have me" (Matthew 26:11).

18 The idea of an afterlife and salvation, which preys on baser feelings of fear and guilt, is an illusion and intended only for the living. If the Church remains helpless to support itself and the faithful in times of great adversity affecting humanity, such as in a pandemic caused by a novel virus, for which there is no credible answer, because prayer does not help, people will be disillusioned and walk away.[14] "Does history support a belief in God? If by God we mean not the creative vitality of nature but a supreme being intelligent and benevolent, the answer must be a reluctant negative."[15]

19 The *summum bonum* for humanity, if it is to achieve success, must be **DEMOCRACY + CAPITALISM + SCIENCE & TECHNOLOGY = PROGRESS & PROSPERITY**

Table 13-1 Differences between Science and Judeo-Christianity

Science	Judeo-Christianity
Reason	Faith
Nature	God
Necessity	Providence
Life	Afterlife
Innovation	Salvation
Evidence	No evidence
Laws/theory	Bible/tradition
Changing	Unchanging
Progressive	Static
Knowledge	Ignorance
Skepticism	Certainty
Pragmatism	Dogmatism
Objective	Subjective
Synergistic	Autonomous

Acknowledgements

Special thanks go to the following people: the editors at Enago and Nick Wilford for their invaluable insights plus advice; Elizabeth Walker and the team at Wordzworth for their technical expertise, without whose respective inputs this book would not have been possible.

Any shortcomings and errors that remain are solely mine.

Notes

Introduction
1. Abelard, "Peter Abelard Quotes."

Prologue
1. Hitchens, "Christopher Hitchens."

Chapter 1 – The Ascent of Man
1. Darwin, The Descent of Man, 398.
2. Ridley, *Genome*, 7–8.
3. Garret-Hattfield, "Animals." Deziel, "Animals."
4. "Hominidae."
5. Stutz, "Embodied Niche Construction."
6. Tuttle, "Human Evolution."
7. The oldest known hominin fossil, of *Sahelanthropus tchadensis*, was discovered in Chad, Central Africa, and it is around seven million years old. It appears to be the oldest human ancestor after what became the genus *Homo* split from what became the genus *Pan*. Additional details on human evolution remain to be determined, including information on the hominin species called the Denisovans, who appeared about 500,000–30,000 years ago and are a complete mystery at the present time (Eldredge, *Extinction and Evolution*, 98; Pavao, "Human Evolution Timeline").
8. Tuttle.
9. Ramirez, "What is the Reason?"
10. Eldredge, 107.

Chapter 2 – The Religions of Man

1. Paul, *Rabindranath Tagore's Gitanjali.*, 39.
2. "Neanderthal Genome Project." Modern genetic studies reveal that interbreeding occurred between *Homo sapiens* and at least two other species, namely, the Neanderthals and the Denisovans "Interbreeding."
3. Diamond, *The Rise and* Fall, 46–48.
4. Hart-Davis, *History*, 20–21.
5. In 1994 an archeological site was discovered in Turkey, called Gobekli Tepe. This is a temple complex built by nomad hunter-gatherers around 10,000 B.C.E., before the appearance of agriculture; this upset the traditional notion of a village settlement formed of a farming community that preceded religious structures. Oliver, "Gobekli Tepe."
6. Older religions of animism and mythology persisted through the Bronze and Iron Ages, although to a much lower degree.
7. Mumford, *The City in History*, 33.

Chapter 3 – In Man's Image and Likeness

1. Xenophanes, "Xenophanes Quotes."
2. Aristotle, "Aristotle Quotes."
3. Spence, *Egyptian Mythology*, 20–21.
4. Durant, *Heroes of History*, 47.
5. Ibid., 45, 47–48.
6. Spence, 22–29.
7. Creative Commons, "Pediments of the Parthenon."
8. Cartwright, "Ancient Greek Religion."

Chapter 4 – Apotheosis

1. Adam, *The Vitality of Platonism*, 125.
2. Sagan, "Cosmos - A Personal Voyage."
3. Wasson, "Alexander the Great."
4. Coogan, Philip, and Freedman, *The Religions Book*, 80–81.
5. "Julius Ceasar."
6. PBS, "Augustus."
7. Valcarcel, "This Roman Emperor."
8. "Nero Considered a God?"
9. Abraham was so obedient that he followed God's order to sacrifice his only son Isaac (Genesis 22:9–10). The ritual of animal or human sacrifice, a common Stone Age practice, is abhorrent.

10. God's promise to Noah never to destroy the earth again after the Flood is usually taken to entail a promise to humanity. Because the story of Noah is in the Torah, it is included in this presentation for completeness' sake.
11. Gilbert and Bacon, *Atlas of Jewish Civilization.*
12. Hasson, "Is the Bible True?"

Chapter 5 – Apotheosis II
1. Libreria Editrice Vaticana, "Why Did the Word?"
2. Kaushik, *Mahatma Gandhi in Cinema,* 137.
3. Matthew 1:1–17 and Luke 3:23–38 have similar presentations of the lineage of Jesus, but there are differences as well. Luke's account includes more generations, reaching all the way back to Adam, while Matthew's account stops at Abraham "Genealogy of Jesus."
4. Human sacrifice in the Iron Age was as revolting then as it is now.
5. Carrier, "Why Gospels are Myth."
6. "Pauline Epistles."
7. Lataster, "Weighing Up the Evidence."
8. "Historicity of Jesus."
9. Schweitzer's expanded edition of *The Quest of the Historical Jesus* was not published in English until 2001. "Albert Schweitzer." "The Quest of the Historical Jesus."

Chapter 6 – The Worship of Man
1. Theophrastus, "Theophrastus Quotes."
2. This quotation attributed to Thomas Jefferson was verified by Ms. Anna Berkes, research librarian, Jefferson Library, Thomas Jefferson Foundation, Inc. on April 13, 2020, through email correspondence Berkes, "Re: research.library.tan."
3. Hodge, "Archibald Alexander Hodge Quotes."
4. Collins and Price, *The Story of Christianity*, 46.
5. Ibid., 59.
6. Mark, "The Medieval Church."
7. Ibid.
8. "Medieval Europe: Church History."
9. Mark, "Religion in Middle Ages."
10. Mark, "The Medieval Church."
11. Creighton-Jobe, *Catholicism*, 60–61.
12. Mark, "The Medieval Church"; Creighton-Jobe, 50–51.

13. Mark, "Religion in Middle Ages."
14. Collins and Price, 113.
15. "Thomas Hobbes."
16. Durant, *The Age of Faith*, 564, 765–766.
17. Ibid., 765.
18. "Medieval Europe: Church History"; Collins and Price, 112.
19. Scholasticism was a method of thinking in which the philosophy of the ancients, such as the logic of Aristotle or Plato, was used to better explain divine revelation and aspects of theology that could not be straightforwardly understood. It began to be taught in 1100 C.E. in the monastic schools that later would evolve into universities. As the environment evolved, so did its method, and it ultimately transformed itself into dialectical reasoning and logic to resolve arguments on secular matters. St. Thomas Aquinas's magnum opus *Summa Theologiae* is considered to be the result of a combination of philosophy (reason) and theology (faith). "Scholasticism."
20. Durant, The Age of Faith, 78.
21. Ibid., 442.
22. Estimates only

Chapter 7 – The Nature of Man

1. Mark, "Protagoras of Abdera."
2. Alexander, "3 Caroline Alexander Quotes."
3. Forster, "E.M. Forster Quotes."
4. Cline, "Renaissance Humanism."
5. A Dominican friar, Johann Tetzel, raised money from the sale of indulgences on behalf of Prince Albert of Brandenburg, who had become indebted to the Fugger banking family, who supplied the money needed to purchase the archbishopric of Mainz. This aroused the ire of Martin Luther. Collins and Price, 132–133.
6. Ibid., 131; Durant, *Heroes of History*, 245–6.
7. "Petrarch."
8. Durant, Heroes of History, 190–1.
9. Ibid., 195.
10. In October 1992, 359 years late, Pope John Paul II apologized on behalf of the Church, calling it an error that Galileo Galilei's science was condemned.
11. Durant, Heroes of History, 319.
12. Francis Bacon, "Sir Francis Bacon Quotes."
13. W, "The Apprentice Art Studio."

Chapter 8 – Eureka!

1. Roger Bacon, "Roger Bacon Quotes."
2. Feynman, "Richard P. Feynman Quotes."
3. "Council of Trent."
4. Durant and Durant, *The Age of Reason*, 238–44.
5. Collins and Price, 147.
6. Durant and Durant, *The Age of Reason*, 245.
7. Snow, *History of the World*, 152.
8. "Mathematics."
9. Hart-Davis, *Science*, 90–91.
10. "William Harvey."
11. Published 1619 C.E. in *The Harmony of the World*, Kepler's Third Law of Planetary Motion shows that the time it takes for two planets to orbit around the sun is in relation to their respective distances from it (Redd, "Johannes Kepler").
12. Included in the *Principia Mathematica* was Isaac Newton's proof of Kepler's second Law of Planetary Motion ("Philosophiæ Naturalis Principia Mathematica").

Chapter 9 – Sapere Aude!

1. Drummond, "William Drummond Quotes."
2. Miller, "Foreword *After the Fall*."
3. Snow, 169.
4. Hart-Davis, *History*, 264.
5. The Dutch built settlements in New Amsterdam (modern New York) in 1624 and in Cape Town, modern South Africa, in 1652.
6. Durant and Durant, *The Age of Reason*, 571–72)
7. Collins and Price, 160–61,164–66.
8. Ibid., 166–69.

Chapter 10 – Prometheus Unbound

1. Confucius, "Confucius Quotes."
2. Sagan, "Carl Sagan's Last Interview."
3. Deism is the philosophical belief in a Supreme Being who created the universe but does not intervene in human affairs. This belief system came to prominence during the seventeenth and eighteenth centuries.
4. Hart-Davis, *History*, 303.

5. Collins and Price, 176.
6. Betro, "The French Revolution."
7. Collins and Price, 177.
8. "First Vatican Council."
9. Collins and Price, 182–83,194–97.
10. Morris, "How Old is Earth?"
11. "Age of the Earth."
12. Hart-Davis, *Science*, 204–5.
13. "Animals."
14. Richard Dawkins asks: "Shouldn't there be some sort of law of decreasing species diversity as we move away from an epicenter—perhaps Mount Ararat?" He narrates the geographical distribution of species in islands and continents found nowhere else, such as the marsupials in Australia, the lemurs in Madagascar, the penguins in the Antarctic, and so on, which only Darwin's theory of evolution can explain (Dawkins, *Greatest Show On Earth*, 268–72).
15. Andrews, "Innovations that Changed History."
16. Hart-Davis, *History*, 292–95; Snow, 214–15.
17. Hart-Davis, *History*, 294.
18. "Timeline of Scientific Discoveries."
19. "Timeline of Historic Inventions."

Chapter 11 – Audeamus!

1. Disareli, "Benjamin Disraeli Quotes."
2. Eliot, "Little Gidding."
3. Armstrong, "Neil Armstrong Quotes."
4. Collins and Price, 178–80.
5. Ibid., 208
6. Ibid., 216–17, 222–25.
7. Hunt, "Lasting Achievements."
8. Klein, "WWI Inventions."
9. Willings, "28 Ways."
10. $E = mc^2$, where E stands for Energy, m for mass, and c^2 for the speed of light multiplied by itself (Falk, "What is Relativity?").
11. Hart-Davis, *Science*, 371.
12. American Chemical Society National Historic Chemical Landmarks, "Penicillin Production."
13. "Internet."

14. "Integrated Circuit."
15. Snow, 312–15.
16. Ibid., 284–85.
17. Ibid., 308–09, 318–19, 322–23.
18. The Hubble–Lemaitre Law states that all galaxies that are moving away from Earth do so at velocities proportional to their distance; the further away they are, the faster they are moving away.
19. "Georges Lemaitre."
20. Snow, 320–21.
21. Ibid., 398–99.
22. Nierenberg, Treu, and Gilman, "Hubble Detects."
23. Hart-Davis, *Science*, 401.
24. Ibid., 400–01.
25. "The Large Hadron Collider."
26. "Cyanobacteria: The Fossil Record."
27. Choi, "7 Theories."
28. Joseph, "Investigating Seafloors and Oceans."
29. Moerkerken, *Seismic Stratigraphy*.
30. Fields, "The Origins of Life."
31. Yong, "The Tree of Life."
32. Munchow, "Maurice Lacroix Masterpiece Mystery and the Solar System."

Chapter 12 – On Morality

1. Einstein, "Albert Einstein Quotes."
2. Bonaparte, "Napoleon Bonaparte Quotes."
3. Law, *The Great Philosophers*, 177–83.
4. Epstein, *Good Without God*.
5. Durant and Durant, *The Lessons of History*, 42.

Chapter 13 – Conclusion

1. Confucius.
2. Sadi-Carnot, "Napoleon LaPlace Anecdote."
3. Durant and Durant, *The Lessons of History*, 43.
4. We could compare the pantheon of Greek gods to the pantheon of superheroes in the contemporary Marvel universe. While previously, the imagined supernatural beings were literally worshiped, today, these modern super-powerful heroes are idolized as part of entertainment (Saunders and Scott, *Marvel Chronicle*).

5. The closest thing to a miracle is stage magic, performed in shows by professional magicians. Yet, they produce only illusions, which are of no practical use whatsoever (Kuhn, "Tricking the Brain").
6. Collins and Price, 210.
7. Durant, *The Age of Faith,* 564.
8. Durant and Durant, *The Lessons of History,* 44–45.
9. "Medieval Life."
10. Collins and Price, 92.
11. A series of experiments on obedience to authority were conducted by Dr. Stanley Milgram at Yale University in the 1960s. He concluded that ordinary people would obey an authority figure who had the signs of legal or moral recognition, no matter the consequence, even to the point of hurting others beyond what would be considered ethical. Milgram summed this up in the following way: "The extreme willingness of adults to go to almost any lengths on the command of an authority constitutes the chief finding of the study and the fact most urgently demanding explanation" (MacLeod, "The Milgram Shock Experiment").
12. Durant and Durant, *The Age of Reason,* 607.
13. Durant and Durant, *the Lessons of History,* 45.
14. A South Korean founder of a fringe sect, Shincheonji Church of Jesus, apologized to his nation for the initial spread of the COVID-19 by the members of his church ("Coronavirus").
15. Durant and Durant, *The Lessons of History,* 46.

Bibliography

Abelard, Peter. n.d. *Peter Abelard Quotes*. Accessed May 08, 2020. https://www.goodreads.com/author/quotes/5490721.Pierre_Ab_lard.

Adam, James. 1911. *The Vitality of Platonism And Other Essays*. Edited by Adela Marion Adam. Cambridge: Cambridge University Press. https://books.google.com.ph/books?id=dNA7AAAAIAAJ&pg=PA125&dq.

n.d. *Age of the Earth*. Accessed April 17, 2020. https://en.wikipedia.org/wiki/Age_of_the_Earth.

n.d. *Albert Schweitzer*. Accessed April 05, 2020. https://en.wikipedia.org/wiki/Albert_Schweitzer.

Alexander, Caroline. n.d. *3 Caroline Alexander Quotes on Lost Gold of the Dark Ages: War, The Worst Journey in the World and The War That Killed Achilles: The True Story of Homer's Iliad and the Trojan War*. Accessed April 02, 2020. https://quotes.pub/caroline-alexander-quotes.

American Chemical Society National Historic Chemical Landmarks. 2008. *Penicillin Production through Deep-tank Fermentation*. June 12. https://www.acs.org/content/acs/en/education/whatischemistry/landmarks/penicillin.html.

Andrews, Evan. 2020. *11 Innovations That Changed History*. March 25. https://www.history.com/news/11-innovations-that-changed-history.

n.d. *Animals*. Accessed April 18, 2020. *https://en.wikipedia.org/wiki/Animal*.

Aristotle. n.d. *Aristotle Quotes*. Accessed March 26, 2020. https://www.azquotes.com/quote/10339.

Armstrong, Neil. n.d. *Neil Armstrong Quotes*. Accessed April 23, 2020. https://www.goodreads.com/author/quotes/331692.Neil_Armstrong.

Bacon, Francis. n.d. *Sir Francis Bacon Quotes on Nature*. Accessed April 03, 2020. https://todayinsci.com/B/Bacon_Francis/BaconFrancis-Nature-Quotations.htm.

Bacon, Roger. n.d. *Roger Bacon Quotes*. Accessed April 04, 2020. https://www.goodreads.com/author/quotes/119129.Roger_Bacon#:~:text=Roger%20Bacon%20quotes%20Showing%201-10%20of%2010&text=.

Berkes, Anna. 2020. *Re: research.library.tan*. Charlottesville, Virginia, April 13. https://mail.yahoo.com/d/search/keyword=monticello/messages/77955?.intl=ph&.lang=en-PH&.partner=none&.src=fp.

Betro, Gemma. 2010. *The French Revolution and the Catholic Church*. December. https://www.historytoday.com/archive/french-revolution-and-catholic-church.

Black, Jonathan. 2014. *The Sacred History : How Angels, Mystics and Higher Intelligence Made Our World*. London: Quercus Editions Ltd.

Bonaparte, Napoleon. n.d. *Napoleon Bonaparte Quotes*. Accessed June 07, 2020. https://www.goodreads.com/quotes/46351-religion-is-excellent-stuff-for-keeping-common-people-quiet-religion.

Bryson, Bill. 2004. *A Short History of Nearly Everything*. New York: Broadway Books.

Carrier, Richard. 2017. *Why the Gospels Are Myth*. November 27. Accessed April 18, 2020. https://www.youtube.com/watch?v=bQmMFQzrEsc.

Cartwright, Mark. 2018. *Ancient Greek Religion*. March 13. https://www.ancient.eu/Greek_Religion/.

Choi, Charles. 2016. *7 Theories on the Origin of Life*. March 24. https://www.livescience.com/13363-7-theories-origin-life.html.

Cline, Austin. 2017. *Renaissance Humanism*. July 16. https://www.learnreligions.com/renaissance-humanism-248119.

Collins, Michael, and Matthew Price. 1999. *The Story of Christianity*. London: Dorling Kindersley Limited.

Confucius. n.d. *Confucius Quotes*. Accessed May 03, 2020. https://www.goodreads.com/quotes/950577-we-have-two-lives-and-the-second-begins-when-we.

—. n.d. *Confucius Quotes*. Accessed April 18, 2020. https://www.goodreads.com/author/quotes/15321.Confucius?page=2.

Coogan, Michael, Neil Philip, and Paul Freedman. 2013. *The Religions Book*. London: Dorling Kindersley Limited.

2020. *Coronavirus: South Korea church leader apologises for virus spread*. March 02. https://www.bbc.com/news/world-asia-51701039.

n.d. *Council of Trent*. Accessed April 06, 2020. https://en.wikipedia.org/wiki/Council_of_Trent.

Creative Commons. 2009. *Pediments of the Parthenon*. August 08. Accessed March 30, 2020. https://creativecommons.org/licenses/by-sa/2.0/deed.en;https://en.wikipedia.org/wiki/Pediments_of_the_Parthenon.

Creighton-Jobe, Reverend Ronald. 2011. *Catholicism*. Leicestershire: Lorenz Books.

n.d. *Cyanobacteria : the Fossil Record*. Accessed April 23, 2020. https://ucmp.berkeley.edu/bacteria/cyanofr.html.

Darwin, Charles. 1874. *The Descent of Man*. September. https://charles-darwin.classic-literature.co.uk/the-descent-of-man/ebook-page-398.asp.

—. 2011. *The Origin of Species*. London: Harper Press.

Dawkins, Richard. 2008. *The God Delusion*. New York: First Mariner Books.

—. 2010. *The Greatest Show On Earth : The Evidence For Evolution*. London: Black Swan Book.

—. 2016. *The Selfish Gene*. Oxford: Oxford University Press.

Dennett, Daniel. 2007. *Breaking the Spell : Religion as A Natural Phenomenon*. London: Penguin Group.

Deziel, Chris. 2018. *Animals That Share Human DNA Sequences*. July 20. https://sciencing.com/animals-share-human-dna-sequences-8628167.html.

Diamond, Jared. 2002. *The Rise and Fall of the Third Chimpanzee : How Our Animal Heritage Affects the Way We Live*. London: Vintage Books.

Disareli, Benjamin. n.d. *Benjanmin Disraeli Quotes*. Accessed April 22, 2020. https://www.azquotes.com/author/4001-Benjamin_Disraeli.

Drummond, William. n.d. *William Drummond Quotes*. Accessed April 12, 2020. https://www.goodreads.com/author/quotes/563311.William_Drummond.

Durant, Will. 2001. *Heroes of History*. New York: Simon and Schuster, Inc.

—. 1950. *The Age of Faith*. New York: Simon and Schuster Inc.

Durant, Will, and Ariel Durant. 1961. *The Age of Reason Begins*. New York: Simon and Schuster, Inc.

—. 2010. *The Lessons of History*. New York: Simon and Schuster Paperbacks.

Einstein, Albert. n.d. *Albert Einstein Quotes About Morality*. Accessed May 07, 2020. https://www.azquotes.com/author/4399-Albert_Einstein/tag/morality.

Eldredge, Niles. 2014. *Extinction and Evolution : What Fossils Reveal About the History of Life*. New York: Firefly Books Ltd.

Eliot, T. S. n.d. *Little Gidding*. http://www.columbia.edu/itc/history/winter/w3206/edit/tseliotlittlegidding.html.

Epstein, Greg. 2010. *Good Without God : What a Billion Nonreligious People Do Believe*. New York: HarperCollins Publishers.

Falk, Dan. 2018. *What is relativity? Einstein's mind-bending theory explained*. April 14. https://www.nbcnews.com/mach/science/what-relativity-einstein-s-mind-bending-theory-explained-ncna865496.

Feynman, Richard P. n.d. *Richard P. Feynman Quotes*. Accessed April 04, 2020. https://www.azquotes.com/quote/920995.

Fields, Helen. 2010. "The Origins of Life." *Smithsonian Magazine*, October. https://www.smithsonianmag.com/science-nature/the-origins-of-life-60437133/.

n.d. *First Vatican Council*. Accessed April 15, 2020. https://en.wikipedia.org/wiki/First_Vatican_Council.

Forster, E. M. n.d. *E. M. Forster Quotes*. Accessed April 02, 2020. https://www.azquotes.com/quote/99644.

Garret-Hattfield, Lori. n.d. *Animals That Share Human DNA Sequences*. https://education.seattlepi.com/animals-share-human-dna-sequences-6693.html.

n.d. *Genealogy of Jesus*. Accessed April 06, 2020. https://en.wikipedia.org/wiki/Genealogy_of_Jesus.

n.d. *Georges Lemaitre*. Accessed April 26, 2020. https://en.wikipedia.org/wiki/Georges_Lema%C3%AEtre.

Gilbert, Martin, and Josephine Bacon. 1998. *The Illustrated Atlas of Jewish Civilization : 3,000 Years of History*. London: Eagle Editions.

Harari, Yuval Noah. 2015. *Sapiens : A Brief History of Humankind.* New York: HarperCollins Publishers.

Harris, Sam. 2004. *The End of Faith : Religion, Terror and the Future of Reason.* New York: W.W. Norton & Company , Inc.

Hart-Davis, Adam. 2015. *History From the Dawn of Civilization to the Present Day.* London: Dorling Kindersley Limited.

—. 2016. *Science the Definitive Visual Guide.* London: Dorling Kindersley Limited.

Hasson, Nir. 2017. "Is the Bible a True Story?" *Haaretz [Tel Aviv, Israel]*, November 01. https://www.haaretz.com/archaeology/MAGAZINE-is-the-bible-a-true-story-latest-archaeological-finds-yield-surprises-1.5626647.

n.d. *Historicity of Jesus.* Accessed March 31, 2020. https://en.wikipedia.org/wiki/Historicity_of_Jesus.

Hitchens, Christopher. 2016. *Christopher Hitchens : 4 Minute Case Against Religion.* July 07. Accessed March 20, 2020. https://www.youtube.com/watch?v=2xt40wSyQjA.

—. 2009. *God Is Not Great : How Religion Poisons Everything.* New York: Twelve Books.

Hodge, Archibald Alexander. n.d. *Archibald Alexander Hodge Quotes.* Accessed April 01, 2020. https://www.azquotes.com/author/20445-Archibald_Alexander_Hodge#:~:text.

n.d. *Hominidae.* Accessed March 23, 2020. https://en.wikipedia.org/wiki/Hominidae.

Hunt, Bill. 2013. *Lasting Achievements of the Second Vatican Council.* June 10. http://theprogressivecatholicvoice.blogspot.com/2013/06/lasting-achievements-of-second-vatican.html.

n.d. *Integrated Circuit.* Accessed April 27, 2020. https://en.wikipedia.org/wiki/Integrated_circuit.

n.d. *Interbreeding Between Archaic and Modern Humans.* Accessed March 29, 2020. https://en.wikipedia.org/wiki/Interbreeding_between_archaic_and_modern_humans.

n.d. *Internet.* Accessed April 27, 2020. https://en.wikipedia.org/wiki/Internet.

Joseph, Dr. Anthony. 2017. *Investigating Seafloors and Oceans : From Mud Volcanoes to Giant Squid.* Amsterdam: Elsevier, Inc. https://www.sciencedirect.com/science/article/pii/B9780128093573000011.

n.d. *Julius Ceasar.* Accessed March 29, 2020. https://en.wikipedia.org/wiki/Julius_Caesar.

Kaushik, Narendra. 2020. *Mahatma Ganhi in Cinema.* Cambridge: Cambridge Scholars Publishing. https://books.google.com.ph/books?id=JkfhDwAAQBAJ&pg=PA137&lpg=PA137&dq.

Klein, Christopher. 2019. *WWI Inventions, From Pilates to Zippers, That We Still Use Today.* March 14. https://www.history.com/news/world-war-i-inventions-pilates-drones-kleenex.

Kuhn, Gustav. 2016. "Tricking The Brain : How Magic Works." *The Conversation*, March 29. https://theconversation.com/tricking-the-brain-how-magic-works-56451.

Langley, Dr. Myrtle, Tony Lane Dr. Jan Bergman, and Andrew Walls. 1994. *The World's Religions.* Oxford: Lion Pubishing PLC.

Lataster, Raphael. 2014. "Did historical Jesus really exist? The evidence just doesn't add up." *Washington Post*, December 18. https://www.washingtonpost.com/posteverything/wp/2014/12/18/did-historical-jesus-exist-the-traditional-evidence-doesnt-hold-up/.

—. 2014. *Weighing up the evidence for the 'Historical Jesus'.* December 15. https://theconversation.com/weighing-up-the-evidence-for-the-historical-jesus-35319.

Law, Stephen. 2013. *The Great Philosophers : the Lives and Ideas of History's Greatest Thinkers.* London: Quercus.

Libreria Editrice Vaticana. 2003. *Why Did the Word Become Flesh?* November 04. https://www.vatican.va/archive/ccc_css/archive/catechism/p122a3p1.htm.

MacLeod, Saul. 2017. "The Milgram Shock Experiment." *Simply Psychology.* https://www.simplypsychology.org/milgram.html.

Mark, Joshua J. 2012. *Protagoras of Abdera: Of All Things Man Is The Measure.* January 18. https://www.ancient.eu/article/61/protagoras-of-abdera-of-all-things-man-is-the-meas/.

Mark, Joshua. 2019. *Religion in the Middle Ages.* June 28. https://www.ancient.eu/article/1411/religion-in-the-middle-ages/.

—. 2019. *The Medieval Church.* June 17. https://www.ancient.eu/Medieval_Church/.

Marriott, Emma. 2016. *The History of the World in Bite-Sized Chunks.* London: Michael O'Mara Books Limited.

n.d. *Mathematics.* Accessed April 05, 2020. https://simple.wikipedia.org/wiki/Mathematics.

n.d. *Medieval Europe : Church History.* Accessed April 01, 2020. https://www.timemaps.com/encyclopedia/medieval-europe-church-history/.

n.d. *Medieval Life.* Accessed May 04, 2020. https://courses.lumenlearning.com/boundless-worldhistory/chapter/medieval-life/.

Milgram, Stanley. 2009. *Obedience to Authority.* New York: First Harper Perennial Modern Thought.

Miller, Arthur. 2012. "Foreword to After the Fall 1964." In *Arthur Miller - Collected Essays*, by Arthur Miller. New York: Penguin Books. https://books.google.com.ph/books?id=SCfbCwAA QBAJ&pg=P T176&lpg=P T176&dq=%E2%80%9C+The+apple+c annot+be+stuck+back+on+the+Tree+of+Knowledge ;+once+we+begin+to+see,+we+are+doomed+and+chal lenged+to+seek+the+strength+to+see+ more,+not+less&source=bl&ots=lKptyZY9.

Mlodinow, Leonard. 2016. *The Upright Thinkers : The Human Journey from Living in Trees to Understanding the Cosmos.* New York: Vintage Books.

Moerkerken, Paul C.H. Veeken and Bruno van. 2014. *Seismic Stratigraphy and Depositional Facies Models.* Amsterdam: Elsevier, Inc. https://www.sciencedirect.com/book/9780124114555/seismic-stratigraphy-and-depositional-facies-models.

Morris, John D. 1995. *How Old Is The Earth According To The Bible?* February 01. https://www.icr.org/article/how-old-earth-according-bible/.

Mumford, Lewis. 1961. *The City in History.* New York: A Harvest Book Harcourt Inc.

Munchow, Joshua. n.d. "Maurice Lacroix Masterpiece Mystery And The Solar System." *Quill and Pad.Com.* Accessed July 09, 2020. https://quillandpad.com/wp-content/uploads/2014/08/Milky-Way1.jpg.

n.d. *Neanderthal Genome Project.* Accessed March 27, 2020. https://en.wikipedia.org/wiki/Neanderthal_genome_project.

Nierenberg, Anna, Tommaso Treu, and Daniel Gilman. 2020. *Hubble Detects Smallest Known Dark Matter Clumps.* January 08. Accessed May 06, 2020. https://hubblesite.org/contents/news-releases/2020/news-2020-05.

Oliver, Mark. 2018. *Built 6,000 Years Before Stonehenge, Gobekli Tepe Is The Oldest Temple In The World.* June 14. https://allthatsinteresting.com/gobekli-tepe.

Paul, Dr. S.K. 2006. *The Complete Poems of Rabindranath Tagore's Gitanjali - Texts And Critical Evaluation.* New Delhi: Sarup & Sons. https://books.google.com.ph/books?id=IproIa_rIv8C&pg=PA39&dq.

n.d. *Pauline Epistles.* Accessed March 30, 2020. https://en.wikipedia.org/wiki/Pauline_epistles.

Pavao, Paul. n.d. "Human Evolution Timeline." *Proof of Evolution.com.* Accessed March 28, 2020. https://www.proof-of-evolution.com/human-evolution-timeline.html.

PBS. n.d. *Augustus.* Accessed March 29, 2020. https://www.pbs.org/empires/romans/empire/augustus_religion.html.

2009. *Pediments of the Pathenon.* Flickr. August 08. https://en.wikipedia.org/wiki/Pediments_of_the_Parthenon;https://creativecommons.org/licenses/by-sa/2.0/deed.en.

2019. *Petrarch.* March 18. https://www.newworldencyclopedia.org/entry/Petrarch.

n.d. *Philosophiæ Naturalis Principia Mathematica.* Accessed April 12, 2020. https://en.wikipedia.org/wiki/Philosophi%C3%A6_Naturalis_Principia_Mathematica.

Ramirez, Israel. 2017. *What is the reason that apes cannot permanently walk on their legs like human can?* March 27. https://qph.fs.quoracdn.net/main-qimg-fb5562a956c24821b3206804a7797152.webp.

Redd, Nola Taylor. 2017. *Johannes Kepler: Unlocking the Secrets of Planetary Motion.* November 20. https://www.space.com/15787-johannes-kepler.html.

Ridley, Matt. 2006. *Genome : The Autobiography of A Species in 23 Chapters.* New York: First Harper Perennial.

Sadi-Carnot. 2019. *Napoleon Laplace Anecdote.* April 09. http://www.eoht.info/page/napoleon+laplace+anecdote.

Sagan, Carl, interview by Charlie Rose. 1996. *Carl Sagan's Last Interview With Charlie Rose* (May 27). https://www.youtube.com/watch?v=U8HEwO-2L4w&t=592s.

—. 1980. *Cosmos - A Personal Voyage*. September 28. https://speakola.com/ideas/carl-sagan-man-in-his-arrogance-1980.

—. 2013. *Cosmos*. New York: Ballantine Books.

—. 1997. *The Demon-Haunted World : Science as A Candle in the Dark*. New York: Ballantine Books.

Saunders, Catherine, and Heather Scott. 2008. *Marvel Chronicle : A Year By Year History*. London: Dorling Kindersley Limited.

n.d. *Scholasticism*. Accessed April 11, 2020. https://en.wikipedia.org/wiki/Scholasticism.

Snow, Peter. 2018. *History of the World Map by Map*. London: Dorling Kindersley Limited.

Spence, Lewis. 1996. *The Illustrated Guide to Egyptian Mythology*. London: Studio Editions.

Stewart, Ian. 2013. *In Pursuit of the Unknown : 17 Equations That Changed the World*. New York: Basic Books.

Stutz, Aaron J. 2014. "Embodied niche construction in the hominin lineage: semiotic structure and sustained attention in human embodied cognition." Edited by Guy Dove. *Frontiers in Psychology - Cognitive Science*. https://www.frontiersin.org/articles/10.3389/fpsyg.2014.00834/full.

n.d. *The Large Hadron Collider*. Accessed April 25, 2020. https://home.cern/science/accelerators/large-hadron-collider.

n.d. *The Quest of the Historical Jesus*. Accessed April 07, 2020. https://en.wikipedia.org/wiki/The_Quest_of_the_Historical_Jesus.

Theophrastus. n.d. *Theophrastus Quotes*. Accessed April 01, 2020. https://www.azquotes.com/quote/1048395.

n.d. *Thomas Hobbes*. Accessed June 07, 2020. https://en.wikipedia.org/wiki/Thomas_Hobbes.

n.d. *Timeline of Historic Inventions*. Accessed April 20, 2020. https://en.wikipedia.org/wiki/Timeline_of_historic_inventions.

n.d. *Timeline of Scientific Discoveries*. Accessed April 19, 2020. https://en.wikipedia.org/wiki/Timeline_of_scientific_discoveries.

Tuttle, Russell Howard. 2020. *Human Evolution.* February 03. https://www.britannica.com/science/human-evolution.

Valcarcel, Jose Antonio Rodriguez. 2020. "This Roman emperor believed he was a god. He was assassinated for it." *National Geographic,* February 26. https://www.nationalgeographic.com/history/magazine/2015/12/roman-emperor-believed-god-assassinated/.

W, Mrs. 2015. *The Apprentice Art Studio: A Classical Approach to Learning From the Masters.* May 05. Accessed June 09, 2020. https://apprenticewiththemastersartclass.blogspot.com/2015/05/renaissance-review-and-leonardo-and.html.

n.d. *Was the Emperor Nero Considered a God?* Accessed March 29, 2020. https://www.reddit.com/r/AskHistorians/comments/90obqq/was_the_emperor_nero_considered_a_god/.

Wasson, Donald. 2016. *Alexander the Great as a God.* July 28. https://www.ancient.eu/article/925/alexander-the-great-as-a-god/.

Wilkinson, Philip. 2018. *The Mythology Book.* London: Dorling Kindersley Limited.

n.d. *William Harvey.* Accessed April 08, 2020. https://en.wikipedia.org/wiki/William_Harvey.

Willings, Adrian. 2019. *28 ways military tech changed our lives.* May 31. https://www.pocket-lint.com/gadgets/news/143526-how-military-tech-changed-our-lives.

Xenophanes. n.d. *Xenophanes Quotes.* Accessed March 26, 2020. https://www.azquotes.com/author/38174-Xenophanes.

Yong, Ed. 2018. *I Contain Multitudes : The Microbes Within Us and A Grander View of Life.* New York: First Ecco.

—. 2016. "Most of the Tree of Life Is A Complex Mystery." *The Atlantic,* April 12. Accessed May 12, 2020. https://www.theatlantic.com/science/archive/2016/04/the-tree-of-life-just-got-a-lot-weirder/477729/.

Zimmer, Carl. 2006. *Evolution : The Triumph of An Idea.* New York: First Harper Perennial.

Index

Note: Page numbers in *italics* refer to illustrations. Page numbers in **bold** refer to tables. Endnote reference include the page number, chapter number (in parentheses), and note number preceded with an 'n'.

A

Abraham (in the Torah), 19, 20, 21, 24, 70, 78(ch4)n9
abstract thinking, 12
Achilles (god), 18
Advancement of Learning (Bacon), 35
Aeneid (Virgil), 18
Africa (Cicero), 33
agriculture, 11
Albert, Prince of Brandenburg, 80(ch7)n5
Alexander, Caroline, 31
Alexander II, 18
Alexander VI, Pope, 38
American Revolution, 50
Ancient Egypt, 14–15
animism, 10, **11**, 66, 78(ch2)n6
anthropomorphism, defined, 15
apes, 4–7, **6**
apotheosis
    Alexander II, 18
    archeological findings, 21
    God of the Jews, 19–20
    Jesus of Nazareth, 24–25
    Jewish people, 20–21
    New Testament gospels, authenticity of, 25
    Paul (apostle) ignorance of, 25–26
    Roman emperors, 18
    Schweitzer, Albert, 26
archeological findings, 21
Aristotle (Greek philosopher), 12, 13, 34, 35
Armstrong, Neil, 55
ascent, of man, 1–7
astronomy, 39–40
atheism, 56
Augustinian, religious order, 38
Augustus (Roman emperor), 18
Australopithecus, 5
Austro-Hungarian Empire, 58
authority, submission to, 28–30

B

Bacon, Francis, 35
Bacon, Roger, 37
Baptists religion, 46
Berkes, Anna, 79(ch5)n9
Berners-Lee, Tim, 58, **61**
Bible
    modern science and, 70–71
    New Testament, 24–27
    Old Testament, 19–21
    printing press and, 34
    Torah, 19, 20, 21, 24
    translations of, 46
Big Bang theory, 59–60
bipedalism, defined, 5
Bonaparte, Napoleon, 50, 66, 69
Book of Days (Ovid), 18
Byzantine Church, 71
Byzantine Empire, 30, 32

C

Caligula (Roman emperor), 18
Candide (Voltaire), 45
Carrier, Richard, 25
Catechism of the Catholic Church, 23
Celsus (Greek philosopher), 28
Charles I, King of England, 44
Christianity, 24, 28–30, 32, 51, 56, 71–72
church, separation from state, 50
Cicero (Roman statesman-philosopher), 33
civil law, 66
Civil War, England (1642-1651), 44
Code of Hammurabi (1754 B.C.E.), 66
Cold War (US and USSR), 58
Collins, Francis, xi
colonization, 59

communism, 56
Confucius (Chinese philosopher), 49, 69
conscience, 66
Constantine (Roman emperor), 28
Constantinople, 32
Copernicus, Nicolaus, 34, 39
Council of Trent (1545-1563), 38
covenants, **19–20**, **25**
creation, science and, 51–52
Crick, Francis, **61**
criminal law, 66
Cromwell, Oliver, 44
customs, 66

D

da Vinci, Leonardo, 32, *36*
Dante Alighieri, 29
dark energy, 60
dark matter, 60
Darwin, Charles, xii, 1, 4, 51–52
David, King, 19, 21, 24
Dawkins, Richard, xi, 82(ch10)n14
Declaration of Independence, US, 50
Declaration of the Rights of Man and the Citizen, France, 50
Deism, 50, 81(ch10)n3
Denisovans (hominin species), 77(ch1)n7
The Descent of Man (Darwin), 4, 52
dialectical reasoning, 80(ch8)n19
Dionysius of Halicarnassus, 18
Disraeli, Benjamin, 55
The Divine Comedy (Dante), 29
DNA (deoxyribonucleic acid), 2, **2–3**, **10**
Dominican, religious order, 28, 38, 80(ch7)n5
Donatello (sculptor), 34

Drummond, William, 43
Dubois, Eugene, 4–7
Durant, Will, 14, 30

E

Eddy, Mary Baker, xi
Edict of Milan, 28
education, 66
Egypt, religion in ancient times, 14–15
Einstein, Albert, 57, 59, **61**, 65
Eldredge, Niles, 6–7
Eliot, T. S., 55
Encyclopedia or a Systematic Dictionary of the Sciences, Arts and Crafts, 35
Enlightenment era, 45, 72
Eucharist, 29, 57
eukaryotes, defined, *62*

F

farming, subsistence, 11
Ferdinand, King of Spain, 33
"Festival of Dangerous Ideas," xi–xii
Feynman, Richard, 37
First Great Awakening, 46
Fleming, Alexander, 58, **61**
forces of nature, 60
Forster, E. M., 31
fossil classification, 5
fossil deposits, locations of, *4*
Francis I, King of France, 32–33
Franciscan, religious order, 28, 38
Franco, Francisco, 56
French Revolution, 50, 66
funeral preparation, 14

G

Galen (Greek physician), 34
Galileo Galilei, 34, 39, 40, **40**, 72, 80(ch7)n10
Gandhi, Mohandas K., 23
genome, 2, *62*
geocentric model, 39
Gilbert, Martin, 21
Gobekli Tepe (archeological site), 78(ch2)n5
gods and goddesses
  in Ancient Egypt, **14–15**
  in Ancient Greece, **16**
  images and likeness of humans, 13–16, 70
  from Judaism, 19–21, 70–71
Golden Rule, 66
gospels, New Testament, authenticity of, 25, 27
Grand Unified Theory, 60
gravity, law of, 40, **41**
Greece
  religion in ancient times, 15–16
  Renaissance era, 32
Groundwork of the Metaphysics of Morals, 66–67
Gutenberg, Johannes, 34

H

Harvey, William, 39
heliocentric model, 39, 40
Heracles (god), 18
Heraclitus (Greek philosopher), 17
Herzog, Ze'ev, 21
Hesiod (Greek poet), 15
historic invention and scientific discoveries, **54**
historical religions, 10–12

Hitchens, Christopher, xi–xii, 71
Hitler, Adolf, 56, 71
Hobbes, Thomas, 35
Hodge, Archibald Alexander, 27
Homer (Greek poet), 15, 33
hominins
    apes versus, 4–7
    fossil deposits, locations of, *4*
    great apes, ancestral link, *3*
    history of, *5*
    skeleton comparison to apes, **6**
Homo erectus, 5, 6
Homo neanderthalensis, 6, 10
Homo sapiens
    emergence from Africa, 6–7
    genera of, *5*
    great apes family, *3*
    mutation of, 10
horticulture, 11
Hubble, Edwin, 59
Hubble-Lemaitre Law, 83(ch11)n18
Human Genome Project, 2
human sacrifices, 24, 78(ch4)n9, 79(ch5)n4
humanism, 33–34, 67
humanist, 31
humans
    DNA compared to Neanderthals, **10**
    DNA compared to other organisms, **2–3**
    settlements and religions, **11–12**
Hume, David, 45, **47**

I

Ignatius of Loyola, 38
Iliad (Homer), 15
images and likeness of in ancient religions, 13–16, 70

Immaculate Conception of Mary, 51
individualism, 46
indulgences, 29, 80(ch7)n5
industrial revolution, 52–53
interbreeding, between *Homo sapiens*, 78(ch2)n2
inventions, 34, 53, **54**, 70
iron, ordeal by, 29
Isaac (in the Torah), 20
Isabella, Queen of Spain, 33
Israelites, 21

J

Jacob (in the Torah), 20
Jefferson, Thomas, 27, 79(ch5)n9
Jesuit, religious order, 38
Jesus of Nazareth, 23, 24–25, 26, 28, 70, 79(ch5)n3
Jewish Bible, 19, 20, 21, 24
Jewish people, 20–21
John Paul II, Pope, 80(ch7)n10
John XXIII, Pope, 57
Josephus (Pharisee historian), 26
Judaism, 19–21, 70–71
Judeo-Christianity, science and, **73**
Julius Caesar, 18

K

Kant, Immanuel, 45, **48**, 66–67
Kelvin, Lord, 51
Kepler, Johannes, 40, **41**, 81(ch8)n12
Kilby, Jack, 58, **61**

L

Lagrange, Joseph, 69
language, development of, 10–11

Laplace, Pierre-Simon, 69
lares (spirits), 18
Large Hardon Collider, 60
Law of Independent Assortment, 52
Law of Segregation, 52
Lemaitre, Georges, 59, **61**
Leo X, Pope, 32–33
liberal theology, 56
life, origin of, 59, 60–61
Locke, John, 35, 45, **46**, 50, 66
Luther, Martin, 32, 34
Lyell, Charles, 51

M

man
    ascent of, 1–7
    images and likeness of in ancient religions, 13–16, 70
    natural rights of, 50
    nature of, 31–36
    religion of, 9–12
    worship of, 27–30
Mathematical Principles of Natural Philosophy (Newton), 40
mathematics, 37, 39
Medici family, 33–34
Mendel, Gregor, 52
Metamorphoses (Ovid), 18
Methodism, 46
Michelangelo, 32
Middle Ages, vital statistics, **30**, 71
Milgram, Stanley, 84(ch13)n11
Milky Way, *63*
Mill, John Stuart, 35
Miller, Arthur, 43
miracles, 71, 83(ch13)n5
monasteries, 29
monotheism, xii, **11**
Montesquieu, Charles, 45, **48**

moral duty, 66–67
morality, 65–67
mores, 66
Moses (in the Torah), 19, 20
motion, laws of, 40, **41**
mummification, 14
mysteries, science and, 59–61
mythology, 11, **11**, 18, 78(ch2)n6

N

nationalism, 59
natural rights of man, 50
nature, forces of, 60
Nazism, 56
Neanderthal Genome Project, 78(ch2)n2
Neanderthals, DNA compared to humans, **10**
Nero (Roman emperor), 18
New Astronomy (Kepler), 40
New Organon (Bacon), 35
New Testament, 24–27
Newton, Isaac, 40, **41**, 81(ch8)n12
"95 Theses" (Luther), 32
Noah's Ark, 52, 79(ch4)n10
nomadic settlements, **11**
Noyce, Robert, 58, **61**

O

Odyssey (Homer), 15
Old Testament, 19–21
On the Fabric of the Human Body (Vesalius), 34
On the Motion of Heart and Blood (Harvey), 39
On the Origin of Species by Means of Natural Selection (Darwin), 52

On the Revolutions of the Celestial Spheres (Copernicus), 34
Oppenheimer, J. Robert, **61**
Orthodox Church, 32
Ottoman Empire, 59
Ottoman Turks, 32
Ovid (*Book of Days*), 18
Ovid (*Metamorphoses*), 18

P

panspermia, 61
papal infallibility, 51
Parthenon, West Pediment model, 15
Paul (apostle) ignorance of, 25–26
Peace of Westphalia (1648), 33, 44–45
penates (gods), 18
Peter, apostle, 28
Petrarch (poet and scholar), 33
philosophers, 18th century, **46–47**
physics, 39
Pietism, 46
Pius IX, Pope, 51
Pius VII, Pope, 50–51
Pius XI, Pope, 56
Planck, Max, **61**
planetary motion, 81(ch8)n11, 81(ch8)n12
Platonic Academy, 34
polytheism, **11**, 14
Presbyterians, 46
Protagoras (Greek philosopher), 31
Protestantism, 38, 44, 46, 57
Ptolemy, 34
purgatory, 29

Q

quantum theory, 60
The Quest of the Historical Jesus (Schweitzer), 26, 79(ch5)n9

R

Raphael (artist and architect), 32
Reformation, 32, 44, 72
relativity, general theory of, 60
relics, sale of, 29
religion
    in Ancient Egypt, 14–15
    in Ancient Greece, 15–16
    beginnings of, 70
    evolution of with human settlements, **11–12**
    history of, 10–12
    of man, 9–12
Renaissance era, 32–35, 71
Renaissance Man, 36
ritual sacrifices, 24, 78(ch4)n9, 79(ch5)n4
Roman Antiquities (Dionysius), 18
Roman Catholic Church
    Catechism of the Catholic Church, 23
    in crisis, 32–33
    in decline, 44–45, 71–72
    in free fall, 50–51
    as international court, 71
    in remission, 56–57
    science and, 34
    in transition, 38
Roman emperors, 18, 28
Rome, Renaissance era, 32–34
Rousseau, Jean-Jacques, 45, **47**, 50
Royal Society (1600 C.E.), 35
Russian Empire, 59
Rutherford, Ernest, **61**

S

Sagan, Carl, 17, 49
Sahelanthropus tchadensis, 77(ch1)n7
scholasticism, 80(ch8)n19
Schweitzer, Albert, 26
science
    creation and, 51–52
    Judeo-Christianity and, **73**
    mysteries, 59–61
    technology and, 57–58
scientific discoveries of 17th century, **40–41**
scientific method, 35
Scientific Revolution, 39–40, 71
scientific studies and 20th century inventions, **61**
scientific studies and historic inventions, **54**
secular ideologies, 72
Semitic peoples, 20–21
separation of church and state, 50
Smith, Adam, 45, **47**, 53, 66
social contract, 50
social norms, 66
socialism, 56
Society of Jesus (Jesuits), 38
Solomon, King, 21
Spanish Civil War (1936-1939), 56
Spanish Inquisition (1478-1834), 38
speech, development of, 10–11
speed of light, 57, 82(ch11)n10
Spencer, Herbert, 35
Spinoza, Baruch, 45, **48**
spirits, belief in, 10
state, separation from church, 50
subsistence farming, 11
Summa Theologiae (Thomas Aquinas), 80(ch8)n19
superstition, 27
sympathy, 66

T

Tacitus (Roman historian), 26
Tagore, Rabindranath, 9
technology, science and, 57–58
Tetzel, Johann, 80(ch7)n5
Theodosius I (Roman emperor), 28
Theogony (Hesiod), 15
Theophrastus (Greek philosopher), 27
Theory of Everything, 60
Theory of Moral Sentiments (Smith), 66
thinking, abstract, 12
Thirty Years War (1618-1648), 44
Thomas, of Aquinas, St., 80(ch8)n19
Thomson, Joseph John, **61**
Torah (Jewish Bible), 19, 20, 21, 24
Tree of Life, *62*
Twain, Mark, xi

U

universe, origin of, 59–60

V

Vatican Council I (1870), 51
Vatican Council II (1962-1965), 57
Vesalius, Andreas, 34
villages/towns, **11**, 70, 78(ch2)n5
Virgil (*Aeneid*), 18
virtue, 66
vital statistics, Middle Ages, **30**
Voltaire (French author-philosopher), 45, **48**
Von Braun, Wernher, 57, **61**

W

water, ordeal by, 29
Watt, James, 53
worship, of man, 27–30

X

Xenophanes of Colophon, 13

Z

Zeus (god), 18

www.ingramcontent.com/pod-product-compliance
Lightning Source LLC
Chambersburg PA
CBHW071006080526
44587CB00015B/2361